VIDEO OBJECT EXTRACTION AND REPRESENTATION
Theory and Applications

THE KLUWER INTERNATIONAL SERIES
IN ENGINEERING AND COMPUTER SCIENCE

VIDEO OBJECT EXTRACTION AND REPRESENTATION
Theory and Applications

by

I-JONG LIN
Hewlett-Packard Laboratories

S.Y. KUNG
Princeton University

SPRINGER SCIENCE+BUSINESS MEDIA, LLC

Library of Congress Cataloging-in-Publication Data

Lin, I-Jong, 1971-
Video object extraction and representation: theory and applications / by I-Jong Lin, S.Y. Kung.
p.cm.— (The Kluwer international series in engineering and computer science; SECS 584)
Includes bibliographical references and index.
ISBN 978-1-4757-8384-1 ISBN 978-0-306-47037-0 (eBook)
DOI 10.1007/978-0-306-47037-0
1. Digital video. 2. MPEG (Video coding standard) 3. Image processing – Digital techniques. 4. Directed graphs. I. Kung, S.Y. (Sun Yuan) II. Title. III. Series.

TK6680.5 .L56 2000
621.388'33—dc21

00-062187

Originally published by Kluwer Academic Publishers in 2000
Softcover reprint of the hardcover 1st edition 2000

Printed on acid-free paper.

For my first mentor,
Professor S.Y. Kung

I-Jong Lin

Contents

Preface ix

Explanation and Index of Outline Pages xi

Acknowledgments xiii

1. INTRODUCTION TO CONTENT-BASED VISUAL PROCESSING 1
   1. Content-Based Information Processing 2
   2. Scope of Book 3
   3. Physical Object vs. Video Object 3
   4. Convergence of Technologies 6
   5. The MPEG Standards 12
   6. Specific Challenges of the MPEG Standards 19
   7. Research Contributions 20
   8. Book Outline 22
   9. Conclusion 22

2. EXISTING TECHNIQUES OF VISUAL PROCESSING 25
   1. Design Implications of Human Vision 26
   2. Image Segmentation 29
   3. Motion Estimation 33
   4. Knowledge Representation 41
   5. Dynamic Programming 43
   6. Compressed Domain 44
   7. Conclusion 45

3. VORONOI ORDERED SPACE 47
   1. Previous Work 47
   2. Notation 48
   3. Definition of Voronoi Ordered Space 48
   4. Properties of Voronoi Ordered Space 51
   5. Definition of the Voronoi Ordered Skeleton (VOS) 55

6. Multiresolution Property of the VOS 57
7. Applications 63
8. Conclusion 64

4. A SYSTEM FOR VIDEO OBJECT SEGMENTATION 67
1. Previous Work 67
2. Problem Simplifications 70
3. Problem Formulation via Surface Optimization 73
4. The Bootstrap Stage 82
5. Surface Optimization 86
6. Results and Analysis 96
7. Conclusion 105

5. ROBUST REPRESENTATION OF SHAPE WITH DAGS 107
1. Previous Work 107
2. Order, Mapping and DAGs 109
3. Partitioning 111
4. DAG-Coding 113
5. Comparing DAG Representations 118
6. Recursive Structure of DAGs 122
7. Conclusion 126

6. A SYSTEM FOR IMAGE/VIDEO OBJECT QUERY BY SHAPE 127
1. Previous Work 127
2. Problem Description 128
3. Shape Extraction via VOS 131
4. VOS Representation via DAG-Ordered Trees (DOTs) 136
5. System Design 140
6. Results and Analysis 148
7. Conclusion 153

7. THE FUTURE OF CONTENT-BASED VIDEO PROCESSING 155
1. Universal Multimedia Access 155
2. MPEG-4 and MPEG-7 System Synergy 158
3. Intelligent Video Transport 158
4. The "New" Media 161

Index 175

Preface

"If you have built castles in the air, your work need not be lost; that is where they should be. Now put the foundations under them." - Henry David Thoreau, *Walden*

Although engineering is a study entrenched firmly in belief of pragmatism, I have always believed its impact need not be limited to pragmatism. Pragmatism is not the boundaries that define engineering, just the (sometimes unforgiving) rules by which we sight our goals.

This book studies two major problems of content-based video processing for a media-based technology: Video Object Plane (VOP) Extraction and Representation, in support of the MPEG-4 and MPEG-7 video standards, respectively. After reviewing relevant image and video processing techniques, we introduce the concept of Voronoi Ordered Spaces for both VOP extraction and representation to integrate shape information into low-level optimization algorithms and to derive robust shape descriptors, respectively. We implement a video object segmentation system with a novel surface optimization scheme that integrates Voronoi Ordered Spaces with existing techniques to balance visual information against predictions of models of *a priori* information. With these VOPs, we have explicit forms of video objects that give users the ability to address and manipulate video content. We outline a general methodology of robust data representation and comparison through the concept of complex partitioning mapped onto Directed Acyclic Graphs (DAGs). We produce a novel, intuitively interfaced image/video query by shape system for our VOPs whose extraction is based upon Voronoi Ordered Space and whose representation and comparison algorithm are based on our work on DAGs. Within a media-based context, this Image/Video Query by Shape is an important functionality in the creation of a "new" media: an intuitive search over video content accessible both locally and over the Internet. In conclusion, we outline the future applications of content-based video processing and link these content-based processing systems synergistically in a proposed MPEG-4/7 hybrid system that uses high-level VOP representations to aid in the low-level VOP extraction.

In other words, my nearly five years at Princeton have been spent searching in the library, pounding code on the keyboard, reading papers, writing papers, or translating theory into code. This book is the readable distillation of the process that will serve as reference or guidepost to those who wish to pursue this area of research.

In other words, my nearly five years at Princeton have been spent on the aspects of human computation that we take for granted – generalization, recognition and segmentation. However, in deference to pragmatism and the coming MPEG-4/7 standards, the main topic of this book focuses on the video object segmentation problem, from an engineering perspective. Although our stated purpose is to support the MPEG-4/7 standards, I hope the connections of this work run deeper. The video object segmentation problem is a solved problem. It is one of the mind's simple preprocessing steps before visual understanding. In fact, it is such an inherent reflex of our visual processing that it must be connected to the structure of the mind and its processes. I hope the system, its algorithms, its analysis, and its results may shed some more light on these mysterious processes that allow us to make sense of this world visually.

What are Outline Pages?

Outline pages are:
- synthesis of presentation slides from conferences and Ph.D. dissertation
- Written for a general engineering audience
- A quick summary of a part of a chapter
- to periodically familiarize and orient the reader to the chapter material
- Formatted like this page:
    - Full Page
    - Outline Box
    - Question as a title
    - Conclusion as an answer to the title

Outline pages are NOT:
- Formal
- Precise
- Implementation-Oriented

Index of Outline pages

pg.5 Why are Video Objects Important?
pg.8 What Technologies Do We Need?
pg.14 How are MPEG Standards Evolving?
pg.27 How do We Best Leverage Current Technologies?
pg.49 How do we use Voronoi Order?
pg.56 How do we use Voronoi Order Skeleton?
pg.69 How are Video Object Segmentation and MPEG-4 Related?
pg.72 How do we Extract Video Objects?
pg.108 How to use DAGs for Robust Recognition?
pg.110 What do we need to apply our DAG approach?
pg.129 Why do we use Shape for Content Query?
pg.138 How do we use DAG-Ordered Trees for Recognition?
pg.160 What do we need for the "New" Media?

Outline Pages are a useful tool in orienting the reader with the text.

Acknowledgments

There are many people without whose support I could not finished this process of graduate school education and subsequent book writing process.

First, I would like to thank Princeton University for such an environment where academic and deep work can flourish. Karen Williams, Sheila Gunning, Stephanie Constanti, John Bittner, Eugene Conover and Jay Plett have provided me all the support and advice. Professors Wayne Wolf, Peter Ramadge, Ruby Lee, Sharad Malik, and Bede Liu all have contributed to my book with their insights and conversations. I would especially like to thank Prof. Ling Guan and Peter Ramadge for their conscientious and careful reading.

The cold hard cash helps also. I would like to thank Mitsubishi Advanced Television Lab for supporting my work in content-based video processing: Huifang Sun for blending their industrial efforts with my thesis work, Anthony Vetro for his expert help in deciphering and translating MPEG documents into English, Ajay Divakaran for deciphering and translating MPEG-7 and advocating our efforts at the MPEG-7 meetings. Their enthusiasm and help made my work that much better.

Having a publisher is also a good thing. Kluwer Academic Publishers have graciously given me this opportunity for my thesis to be something more than three unread copies in a Princeton Library. Many thanks goes to C. Anne Murray for getting the details correct and Jennifer Evans for guiding me through the publication process.

As I've noted in my previous incarnation at Stanford, what I remember most from a school is the students. The diversity and character of Princeton students creates a creative and pleasant atmosphere of learning. First, in expressing my frustration in the thesis, I have played a goodly amount of basketball and I would like to thank the morning basketball crew. It's hard to find a good pick-up basketball game and now, in retrospect, I ask for their forgiveness for any bruises or injuries (mental and/or physical) that they might have sustained by playing me: Dave Moffett, Anthony Bishop, Tom Carroll, Richard Sarpong, Jeremiah Helms, Susan Rea, Andrea Thies, and, of course, John (Beta) Dudek. I would like to thank the computer engineering graduate students at Princeton for their camaraderie and *esprit de corps*: Zhao Wu, Yang-Bing Li, Jason Fritts, David Rhodes, and, of course, Kamal Khouri. Although I am probably a little overly independent, my research group was very supportive: Yen-Kuang Chen, whose work ethic always shamed into doing a little more, and Yun-Ting Lin, who always gave her support with a measure of sweetness. When finishing up my thesis, I became a surrogate lab member of the Schutt Lab in Chemistry and Schutt Lab provided, as Hemingway would say, a clean and well-lit place for my last few months. To me, taking in a graduate student has always been a gesture of kindness, so I must thank Amy Cocina, Ilana and Greg Bowman, Kartik Narayan, Rebecca Page, Don Huddler, and, of course, Prof. Clarence Ernest Schutt. Finally, it has been my pleasure to work with Robert Dick, who is far and away a better engineer than I ever could be and has imbued me with a more liberatarian view and greater appreciation for the free software movement, Malcolm Carroll, whose appreciation of well-thought idea always made for a enlightening, fun conversation and timely support, and Constantine Kreatsoulas, who shares the same type of bizarre connectivist intellectual framework and made my thought processes a little less isolating.

My family has always been support through my life and this graduate career is just one more thing in whole continuum of trials and tribulations. My parents were particularly sympathetic to the low points in my graduate career and helped me through the darker times. Their love of science has become a better part of who I am.

Finally, I would like to thank my advisor, Prof. S. Y. Kung. Issues of intellectual aesthetics aside, Prof. Kung has been my first mentor in educational process. For a person like myself who has a tendency to be overly independent, Prof. Kung has been unusually understanding with me, providing the berth that I require while teaching me the finer points of scientific writing, research and critical thinking in his own unique way. My four and a half years in graduate school have been a continual learning process due to his dedication and tenacity to express knowledge in the clearest vision and with the clearest language possible. I am deeply grateful for his efforts and hope, in return, that his experience with me has taught him a thing or two.

Chapter 1

INTRODUCTION TO CONTENT-BASED VISUAL PROCESSING

Energy and production now tend to fuse with information and learning. Marketing and consumption tend to become one with learning, enlightenment and the intake of information. This is all part of the electric implosion that now follows or succeeds the centuries of explosion and increasing specialism. The electronic age is literally one of illumination. Just as light is at once energy and information, so electric automation unites production, consumption and learning in an inextricable process.
- Marshall McLuhan, *Understanding Media, The Extensions of Man*

The Information Age is upon us. The raw material of this age is information; the product is content. The computational might of the integrated silicon transistor drives this age forward. In the eighties and nineties, the computer industry infrastructure was geared to produce faster, more powerful and cheaper engines on these pieces of silicon. Already, we use computers everyday: explicitly, to write programs, to create spreadsheets, or to make airline reservations with workstations and personal computers; implicitly, to control our household appliances, telephones, and cars with microcontrollers and embedded systems. However, the role of computer as a controller is merely the first step in the coming information age and the specialized action of "using a computer" shows how immature this information age is. The electrical power infrastructure is mature when and only when one can plug into a wall socket and not have to understand the concept of the step-down transformer. The next great step of this information age is to apply this computational power into systems that support a dynamic "new" media based upon audio-video content-based technologies. These technologies not only compress and transport but also allow users to actively transform and create content. The information age will not be defined by deliverable products such as a 1GHz microprocessor, but by computational systems. In short, the software systems must evolve along with the hardware. These computational systems describe a dynamic system that encompasses not only ubiquitous viewing and trading of video content but also manipulation and accessibility to previous work and the creation of new content. These systems will ultimately transform multimedia content into a novel form of dynamic communication, a language based on these digital technologies.

Within the information age, this book concentrates upon a single subsection, visual (both image and video) processing and, within visual processing, this book concentrates upon two processes: video object extraction and video object representation. Although we only treat two processes in the whole field of video processing, we hope to show the dynamic of this information age. In the following sections, we show how our work and this book play a crucial role in the information age. By relating extraction and representation, we also hope to hint at the future technologies that will drive the information age.

1. CONTENT-BASED INFORMATION PROCESSING

To reflect the challenges of this Information Age, the focus of signal processing is shifting toward content-based information processing. In the past, the traditional signal processing primarily dealt with reproduction and communication, i.e., the quality of work was always self-referential to the original information. This type of work was mathematically rigorous, since an objective comparison could always be made to the original. By definition, content is subjective and content-based information processing defies complete formalization. Although subjective, content-based analysis must always be compared in relation to how we as human beings process content.

The study of content-based processing requires an understanding of the mechanics and system dynamics of content-based information processing. A deep understanding of content-based information processing relies on the identification of the intermediate representations that content-based information processing uses. We need to understand these intermediate representations in order to:

1. break down the content-based information processing into tractable subprocesses,

2. have results that will fit into the current informational infrastructure of human processing,

3. and have results that are useful to future content-based information processing.

Instead of treating content-based information processing as a monolithic problem, we break the task as a series of more manageable problems, transformations from one intermediate representation to the next. A solution for one of the transformations gives the user a tool to explicitly manipulate the content and works toward the completion of the whole content-based information processing task. In this book, we wish to duplicate our own visual biological preprocessing in explicit terms for video processing. In visual processing, we study the intermediate representations for both 1) an object-oriented extraction process as technologies for content-based analysis and 2) an intuitive similarity criterion of shapes.

2. SCOPE OF BOOK

Concentrating on video processing, we now define the scope of this book work, in both theory and practice.

From a theoretical perspective, we only deal with the visual projections that we see rather than the physical object itself. For instance, in our video object extraction algorithm, we are only concerned with extracting the areas of the single frame of video onto which a physical object projects, without considering its physical parameters such as its 3-D structure or size. We do not cover physical object reconstruction in this book and we only cover a limited subset of computer vision problems. Although some problems that deal with visual projection of objects cannot be solved unless we consider the physical object itself, these limitations will be clearly defined and serve to distinguish between levels of content-based processing.

From a pragmatic perspective, this book is looking for robust tools for extraction, comparison, and manipulation of video objects. We concentrate on two key functionalities: extraction and representation. In both cases, since the circumstances and user demands can be so varied, we must commit to a limited subset of functionality in order to build complete systems. In the future, we envision a framework of tools to meet user demands. Although we focus upon specific systems, we can learn much of the general design methodology for these content-based processing systems.

3. PHYSICAL OBJECT VS. VIDEO OBJECT

To clarify the objectives of this book, we define and differentiate two key terms: the *physical object* and the *video object*. The concept of a

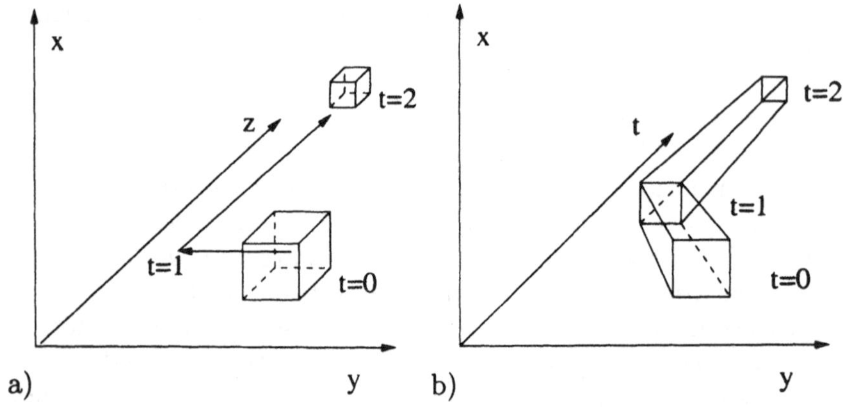

Figure 1.1. Physical Object vs. Video Object: a) a cube as a Physical Object moving through time in the real world, translating with constant velocity in the y direction from t=0 to t=1 and then translating in the z direction from t=1 to t=2 and b) its corresponding Video Object through a projection onto the x-y plane. Note that the time correspondence between two spaces.

physical object is a philosophical postulate that frames the book. The video object is a simple result of this postulate, and a simpler, intermediate representation of the physical object.

We now define the fundamental elements of this book. We define the *physical world* as the domain where the ground truth exists, the real 4-D physical space $(x_{phys}, y_{phys}, z_{phys}, t_{phys})$ (3-D space with time component). Three dimensional objects that move around in physical space (see Fig 1.1a) are called *physical objects.* Our own view of this world is limited: from a single camera view, the physical world is projected onto a 3-D space (x, y, t) that we call the *video space.* The *video sequence*, $I(x, y, t)$, is a real-valued function over the video space that represents a series of grayscale images that we see. A timeslice of a video sequence $(I(x, y, t)|_{t=t_0})$ is called a frame t_0 in this book. For a given time instant, each 3-D object in the physical world is projected as a flat 2-D image. Moving through the physical world, a 3-D object is projected as a time-indexed series of 2-D images. The X and Y dimensions of the space are symmetric. Along the time dimension, there exists a strong correspondence between images along the X-Y planes due to the fact that physical objects either are stationary or must move away, but do not suddenly appear and disappear.

POSTULATE 1.1 (OBJECT-ORIENTED DECOMPOSITION) *The video sequence $I(x, y, t)$ is a composition of projections of a number of physical objects that exist in the physical world. The video sequence may be parti-*

Why Are Video Objects Important?

Physical Object = objects that we are accustomed to manipulate in 3-D space

Video Object = a 3-D spatio-temporal region that contains the projection of a physical object onto a video sequence

The objective of this book is to describe technologies that enable users to 1) extract video objects from video sequence and 2) search these video object w.r.t. their content.

Other fields such as Computer Vision treat Video Objects as an intermediate step to implement a particular functionality. This book is centered on Video Objects as the basis for understanding and manipulating video content.

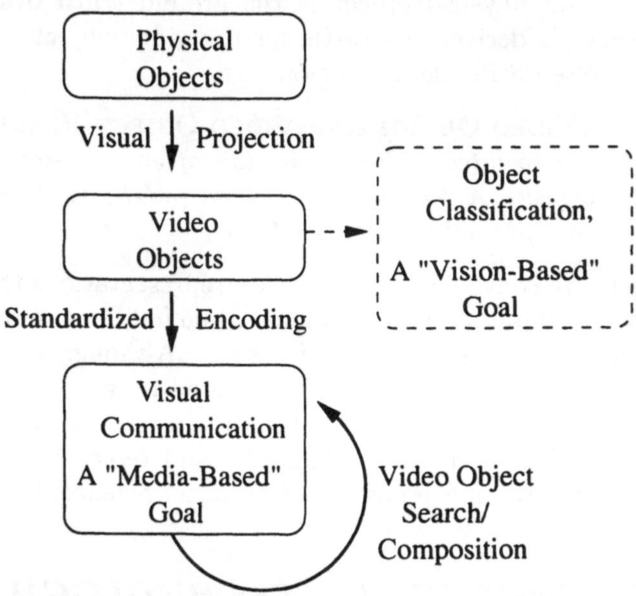

We treat Video Objects are the basic elements in a new media of visual communication

tioned into volumes onto which each physical object projects. The value of $I(x, y, t)$ over each volume is the projection of one and only one object.

Deriving physical objects from the video sequence is the true aim of computer vision and is a notoriously difficult problem. The representation of the physical object is the three dimensional representation that matches the physical ground truth. From the video sequence, we must not only differentiate between physical objects, but also consider how these physical objects project onto the video space. However, what we see is merely the projection of the physical world and much information about the physical object is lost through the process of projection. For instance, in Fig. 1.1b, any number of objects with a square face could have created the same projection. If we consider only the projection of the physical objects in the video space, this restriction vastly simplifies the problem by **NOT** dealing two major issues in computer vision:

1. Reconstruction of the 3-D physical structure
2. Descriptions of how 3-D physical structure projects onto video space, e.g., camera and perspective models.

Although the 3-D physical object is the ground truth from which the video sequence is derived, we settle for the video object, a simpler intermediate representation defined below.

DEFINITION 1.1 (VIDEO OBJECT AND VIDEO OBJECT PLANE) *The intensity function associated with a projection of one physical object is called its Video Object. A Video Object Plane (VOP) is the subset of the Video object that lies within a single frame.*

Video Objects are convenient intermediate representations that avoid the complexity of computer vision problems while fulfilling an important functionality in content-based video processing. Although video object extraction and representation may be considered only a preprocessing step for the determination of a physical object, this preprocessing is still difficult and our solutions require techniques and research from many different areas of Electrical Engineering, Computer Science, Biology and Psychology.

4. CONVERGENCE OF TECHNOLOGIES

The complex nature of content-based processing necessitates the convergence of many fields due to the mixed nature of the problem (as shown in Figure 1.2): high-level and low-level analysis, image and temporal information extraction, representation and extraction, system design, algorithmic design, computational engines, and mathematics.

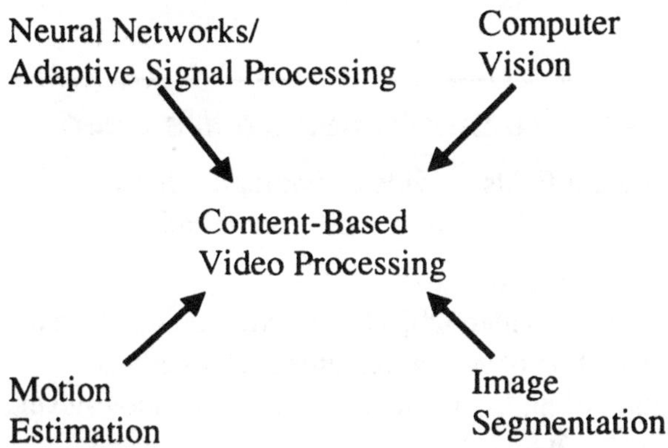

Figure 1.2. Convergence of Technologies for Content-Based Video Processing: Image Segmentation and Motion Estimation provide numerous means of extract and organizing explicit visual information; Computer Vision, Neural Networks and Adaptive Signal Processing provide the computational and algorithmic framework that fits the visual information to coherent representation of video objects.

IMAGE SEGMENTATION

Image segmentation can be considered a special case of our video object extraction upon a single frame (see Figure 1.3). The image segmentation techniques provide many methods of feature extraction and integration of contextual information that form the starting point for most of work in our book. Image segmentation must extract, categorize, arrange and connect visual features with contextual information to find a coherent segmentation. We leverage these technologies and techniques into our video object extraction and representation systems and extend them to handle multiple frames.

Image segmentation covers both the use of spatial correlation and propagating high-level information into segmentation. Image segmentation provides us with many useful low-level feature extraction tools: the concept of locality, spatial correlation and edge detection. Topics such as color and texture identification are also relevant to this book.

Image segmentation must derive its understanding from only a single frame of information; therefore, representations of *a priori* video object knowledge must provide much of the segmentation information. Although we avoid the high-level understanding algorithms that are limited by their specificity, in our system design, we use many image segmentation algorithms that use global information without loss of generality such as edge joining algorithms, region growing, hierarchical representation and clustering algorithms. In video processing, we have the benefit of working with multiple frames and can make use of both spatial and

What Technologies Do We Need?

Many different fields provide technologies that we require for Video Object Extraction and Representation:

Image Segmentation
- special case of Video Object Extraction on one frame
- basic extraction of visual features such as edges
- integration of high-level understanding with low-level features
- similar problem formulation to our video object extraction

Motion Estimation
- a low-level feature that contains high-level information
- key feature that localizes and detects objects
- an alternate description of the video sequence
- accounts for the majority of the temporal information

Neural Networks/ Adaptive Signal Processing
- computational framework for our vision problems
- acts as both representational and computational framework
- adapts well to changing visual conditions and content
- evaluation of content-based systems are similar to recognition problems

Computer Vision
- use similiar technologies as with computer vision
- although using same technologies, our goals are for the creation of a new media, rather than any direct functionality
- in a sense, we use only a subset of computer vision technologies
- existing computer vision systems serve as examples of translating biological function onto digital implementation

Design of Content-Based Analysis Sytems Require Combinations and Hybridizations of these State-of-the-Art Technologies.

a)  b) 

Figure 1.3. Image Segmentation: image segmentation problems often require a contextual knowledge to create a coherent segmentation: for instance, a) is there an embedded triangle or a coincidental placement of semi-circles? b) could you find the dalmation in the picture, if you didn't know there was a dalmation?

a)  b) 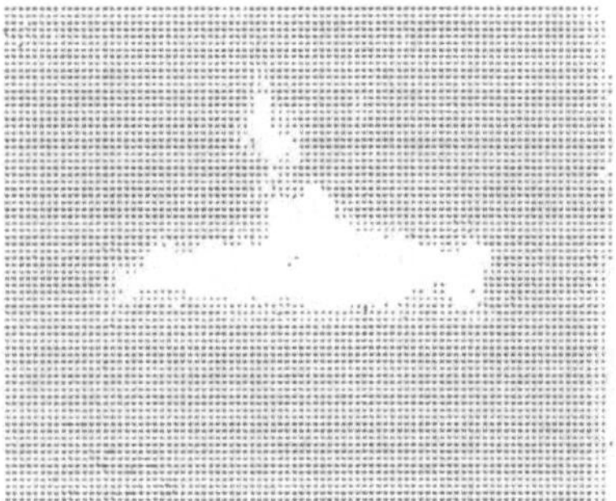

Figure 1.4. Motion Estimation As Source of Object Membership Information: a) a frame of a coastguard sequence, a non-trivial image segmentation problem b) the dense motion field from motion estimation techniques, revealing the moving coastguard ship from the background

temporal correlation. Instead of high-level image understanding such as [Binford, 1982], we will depend on motion, a information source unavailable to image segmentation, for our high-level analyses such as object detection and localization.

MOTION ESTIMATION

Motion estimation not only provides insight into object membership (see Figure 1.4), but also provides valuable algorithmic techniques. While image segmentation techniques are derived from spatial correlation, motion estimation extracts an another set of features derived from spatiotemporal correlation [Lee and Blake, 1999]. Motion estimation derives the seemingly low-level features of projected motion from the video sequence that our systems use. However, projected motion often implies high-level information such as object membership. For our algorithmic design, motion estimation exemplifies a video processing tool that merges both high-level understanding and low-level computation.

Throughout this book, we use estimates of projected motion to robustly detect and localize objects. Motion estimation plays a crucial role in our systems because it provides a robust discriminant to find object boundaries and the informational bootstrap step in our video object extraction system. Instead of a simple function $I(x, y, t)$, motion estimation describes the video sequence as set of moving objects and forms an alternate description of the video sequence as sets of pixels that are correlated through time.

Our algorithmic work in our systems has much of same structure as traditional motion estimation algorithms. Motion estimation algorithms are themselves a type of content-based processing, since projected motion often can only be resolved when content is taken into account. The computation of motion field in Section 2.7 uses many of the same techniques of our video object extraction in Chapter 4 such as its energy function formulation, its counterbalanced optimization and its iterative updating scheme.

NEURAL NETWORKS/ADAPTIVE SIGNAL PROCESSING

Although the two previous sections have concentrated upon deriving visual artifacts, comprehension of the video sequence depends upon the unseen (*a priori* knowledge) as much as the seen (the video sequence, $I(x, y, t)$). Neural networks and adaptive signal processing form the core algorithmic technology for robust processing of spatial and temporal features, since they have the key properties of content-based processing systems:

1. adaptability to the content within a given data set,
2. the ability to integrate high-level direction into low-level processing by their connectivist computation,
3. and the ability to adjust to the long-term definition of content through learning.

Within the context of our video processing tasks, neural networks and adaptive signal processing provide a good solution to content-based analysis problems since they closely mimic human computational techniques.

The ability to learn and adapt contextually are strengths of neural computation and form the algorithmic basis of our content-based information processing.

Our adaptive solutions are merely variations upon, or extensions of, previous neural network applications in different fields such as image processing, speech recognition, optical character recognition and predictive coding. For instance, our surface optimization algorithm in Chapter 4 was inspired by the Hidden Markov Model training algorithms used in speech recognition [Rabiner, 1989]. Our graph comparison algorithm in Chapter 6 is based on our work on cursive handwriting recognition. The major contribution of this book is extending the system design methodology that uses neural networks rather than the technology of neural networks itself. Content-based information processing often is low-level processing that integrates a high-level feedback mechanism into analysis. Our solutions for video object extraction and representation are examples of novel system designs based on adaptive signal and neural network technologies.

COMPUTER VISION

Although we share many of the computer vision technologies, the goal of this book is not to mimic the functionalities human visual system, but rather to extract and manipulate the visual representations that the human mind can understand, manipulate and use in communication. We call these two different goals, *a vision-based* goal and *a media-based* goal. This book is geared toward a media-based goal.

Toward a vision-based goal, this book could be considered a minor field in computer vision. Two main fields of computer vision are already beyond the scope of this book: control and feedback to the data acquisition level (active vision) and the use of 3-D reconstruction techniques (object reconstruction). Our main objectives of this book are "only" precursors of computer vision. As mentioned in Section 3., we will deal with the extraction and representation of video object, not the physical object itself.

Toward a media-based goal, computer vision plays a supporting role to our work as implementation technologies and theoretical background. Computer simulation and study of these biological systems give insight into the vision process. Motion analysis and image segmentation may be considered to be a part of computer vision. Much of computer vision has been devoted to object and scene representation [Ballard and Brown, 1982]. In the next section about video standards, we further distinguish the computer vision from content-based analysis. While video standards are a tangential subject for computer vision, they are a key technology in

Standard	Focus	Functionalities
MPEG-1/2	Compression	Storage, Transmission, Bandwidth Utilization: "The first standard, MPEG-1, was the coding of combined audio-visual signal at a bit rate around 1.5 Mb/s. ... MPEG-2 has been a very successful standard, pieces of equipment that claim conformance to it have been produced by the millions, receivers for digital satellite broadcasting being the most popular. " [Chiariglione, 1997]
MPEG-4	Extraction and Synthesis	Compressed Domain Processing: "at the heart of the so-called content-based MPEG-4 Video functionalities is the support for the separate encoding and decoding of content (i.e., physical objects in a scene). ... This MPEG-4 feature provides the most elementary mechanism for interactivity and manipulation with/of content ... in the compressed domain." [Sikora, 1997]
MPEG-7	Representation and Search	Support for large-scale video object libraries: "MPEG-7 will specify a standard set of descriptors that can be used to describe various types of multimedia information ... To allow for fast and efficient search for materials of a user's interest." [Group, 1999]

Figure 1.5. Summary of MPEG standards and their movement toward content-based representation.

content-based processing. Our work is focused on visual communication and video standards provide the language we require for communication.

5. THE MPEG STANDARDS

Now that we have outlined our technologies, we present the language of content-based video processing, the standards of Motion Picture Experts' Group (MPEG) [Chiariglione, 1997] that support the media-based goals of content-based video processing.

In 1988, the MPEG committee was established to promote a standardization of video data. The MPEG committee neither builds systems nor writes code, but proposes universal standards for expressing video content (See Figure 1.5). By limiting the scope of MPEG to syntactic level,

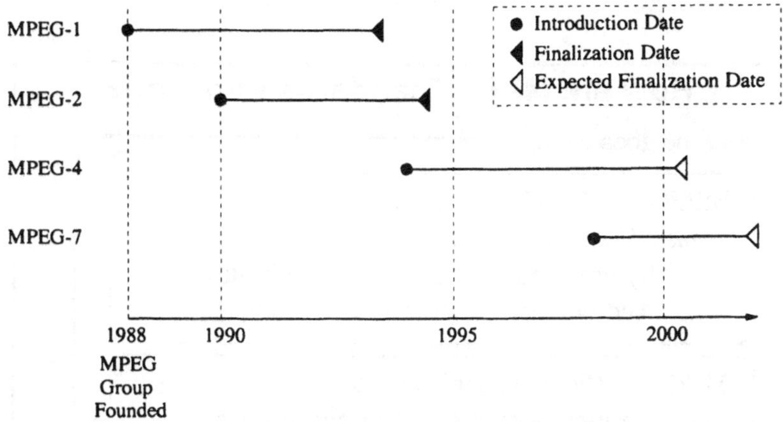

Figure 1.6. Timeline of MPEG Efforts: 1988, Establishment of MPEG and Start of MPEG-1; 1990, Start of MPEG-2; Aug. 1993, MPEG-1 Standard Finalized ; 1994, Start of MPEG-4 Project ; Nov. 1994, MPEG-2 Standard Finalized ; 1998, Start of MPEG-7 Project ; Oct. 1998, MPEG-4 Standard Finalized ; Sept. 2001 , Projected Date of Finalization of MPEG-7 Standard.

MPEG meetings provide a forum for academic and industrial interests to conceptually develop video standards. The discussions at the syntactic level frame problems and their implications upon the nature of the video information representation and extraction. Important issues may be introduced in theoretical fashion at the syntactic level while its relevance may be discussed in terms of the extending the syntax. The MPEG standards also aid implementation by streamlining application and hardware interface. The MPEG meetings provide a regularly timed meetings to compare results, organize current efforts, and steer research toward the future challenges of video processing (see Figure 1.6).

In our own work, the MPEG standards guide our research efforts. Just as the MPEG standards help to focus the effort of industrial development, they also provide support and motivation for the futuristic applications and fit research efforts within the larger context of content-based video processing. Instead of working on a single content-based processing problem in isolation, the MPEG standards provide the context within which we can understand 1) what input the system receives, 2) how to evaluate the system results and 3) how the system must interface to other systems. Each major video standard (MPEG-1/2, MPEG-4, MPEG-7) reflects the structure of this book in theory and application. Although the MPEG-1/2 standard is tangential to our work, it introduces and establishes important concepts about the video sequence. The MPEG-4 standard introduces our first key content-based problem of video object extraction. The MPEG-7 standard introduces our second key content-

How are MPEG Standards Evolving?

As time goes on ...

MPEG-1 (For Content Compression)
- Basic Framework of Video Processing:
  Syntax, Source, Delivery, Demultiplexer
  Audio/Video Compression Techniques

MPEG-2 (For Content Transport)
-Generalization of MPEG-1 for transport over Bandwidth-limited Channels
-Industry Standard used in Satellite TV and DVDs

MPEG-4 (For Content Packaging)
- Provides A Syntax to support:
  - Video Object Segmentation
  - access to individually compressed Video Objects
  - Video as a composition of Video Objects

MPEG-7 (For Content Indexing and Search)
- Provides A Syntax for:
  - Video Object Representation:
  - Descripion Schemes (DS) to search and browse through archives of Video Objects to find desired video content

MPEG-21 (For Content Distribution)
Technologies for e-marketplace: intellectual property protection, electronic commerce, semantic representation, ... ???

The trend is toward more content description and more content-based analysis to make video content as accessible as possible.

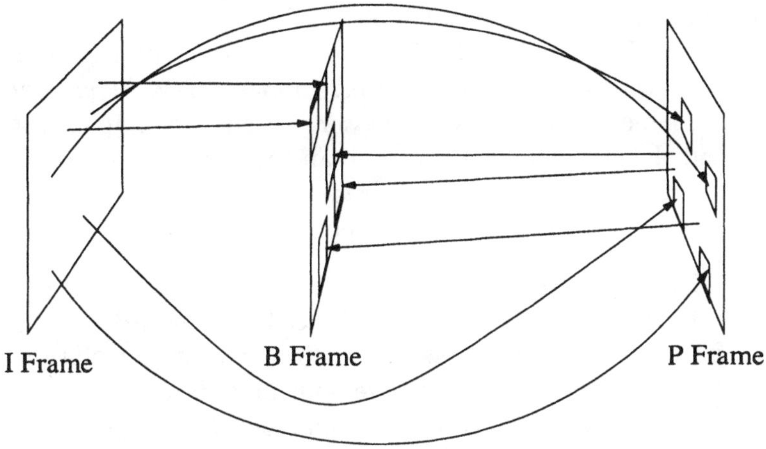

Figure 1.7. MPEG-1/2 Conceptualization of the video sequence, a series of frames described and classified as I (no references to other frames), P (references to a previous frame in the video sequence), and B (references to previous and next frame in the video sequence) by their use of temporal correlation.

based problem of video object representation. In the next sections, we will summarize the relevant points of each standard and connect each standard to the chapters in this book.

MPEG-1/2

In 1988, the MPEG committee introduced MPEG-1 standard as a video compression standard for fixed storage. Anticipating the Internet growth, the MPEG-2 standard was launched to support video over bandwidth limited channels such as broadcast, satellite or Internet. These standards have also introduced many key concepts are the *de facto* basic vocabulary of video coding. The MPEG-1/2 standards [Bhaskaran and Konstantinides, 1995] support following elemental concepts of the video/audio data representation (See Figure 1.7):

1. Frequency-based intra-frame compression
   From the JPEG standard [Wallace, 1991], MPEG committee chose Discrete Cosine Transform (DCT) coding since the DCT coding can appropriate more bits for lower-frequency components of the signal. Psychovisually, smooth images are more visually appealing and less visual content is lost. In addition to integrating human perception into analysis, the bias toward allocating more bits for lower frequency components also justifies a multiresolution representation w.r.t. frequency subsampling.

2. Data Granularity (Block Size)
   For coding efficiency, the MPEG-1/2 standards set the basic unit of video information, an 8x8 or 16x16 block pixels (*a macroblock*). Although some of these constraints were based upon implementation-dependent details such as buffer-size and microprocessor datawidth, the optimal data size granularity reflects the effect of spatial locality/correlation, independent of content.

3. Motion Estimation (I/P/B frames)
   For images within a sequence, the MPEG-1/2 standards classifies the image frames into three classes by their temporal correlation: I, P, and B frames (short for Image, Previous and Bi-directional, respectively). The encoder can describe a macroblock with reference to a macroblock in a frame that has already been decoded, instead of its DCT coding, to reduce bit-rate. I frames have no references to other frames and are completely intra-coded; P frames uses macroblock references from a previous (in the time of the video sequence) frame that have already been decoded; B frames uses both previous and next decoded frames. Although 3-D DCTs compressed as well as a scheme with these motion references, MPEG-1/2 standards chose the concept of P,B frames because of their simplicity [Bhaskaran and Konstantinides, 1995]. This method of coding also justifies our motion-based representation as an alternate representation of the video sequence.

4. Rate Control
   Rate control is an encoder-side technology that allows the encoder to adapt its bit-rate to changing network conditions for best quality of video and introduces feedback mechanisms and content-based quality into the working vocabulary. In bandwidth limited environments, rate control also implies that users prefer quality of content over reproduction in their video playback.

For our own work, the MPEG-1/2 standard conceptually defines our intuitive representation for our video sequence, a series of frames with temporal references that use still image compression techniques for intra-frame coding. The MPEG-1/2 standards are focused toward compression, independent of the content. The MPEG-4 standard drastically changes our conception of the video sequence toward content-based processing.

MPEG-4

The MPEG-4 standard is our reference point in our work on video object extraction. While the MPEG-1/2 standards are primarily coding

Figure 1.8. Object Addressability in the MPEG-4 standard: Instead of coding as a single stream, the MPEG-4 standard supports an object-addressable syntax, i.e. a syntax that allows the content of the video sequence to be addressed by object.

efforts for compression, the MPEG-4 standard is an *object-addressable* video coding syntax, as shown in Figure 1.8. The MPEG-4 standard is a quantum leap because its syntax recognizes the implication of existence of physical objects on the video sequence. Unlike the MPEG-1/2 standards, the MPEG-4 standard is mainly concerned with the extraction and synthesis of video content, addressing spatio-temporal blocks of video information by their object membership. The MPEG-4 standard explicitly supports the concept of a video object. The syntactic support of video objects implies a new type of video processing that is beyond compression. The MPEG-4 standard states that, within a video sequence, there exists a number of video objects that compose the video sequence (as mentioned in Section 3.). In the MPEG-4 standard, the video sequence is no longer the ground truth of the video sequence and is merely the projections of physical objects. While MPEG-1/2 standards were content-independent, the MPEG-4 standard is concerned with not only the video sequence but also the implications of the physical object(s).

The MPEG-4 standard defines one of the major challenges of this book: *video object segmentation*, i.e., the partitioning of video sequence into video objects. When we discuss the video object segmentation prob-

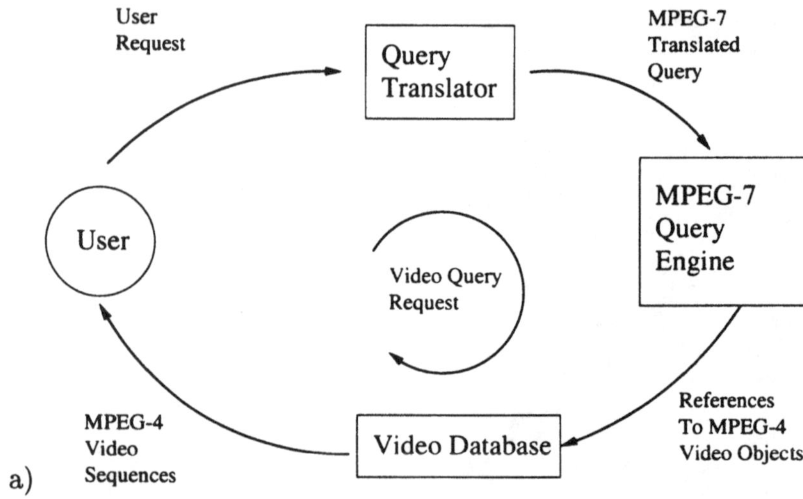

Figure 1.9. MPEG-7 role as the language of video content: starting at the user, a request for video content is issued. The query is then translated into an intermediate MPEG-7 form and a search of similar MPEG-7 representations of video content is initiated. Similar MPEG-7 representations of objects are found and their corresponding video objects are sent back.

lem in Chapter 4, we refer to the MPEG-4 standard for our test set, application specifications, and ground truth results for comparison.

MPEG-7

Just as the MPEG-4 standard provided focus for extraction and synthesis functionalities, the MPEG-7 standard explores issues of video object representation and search. While the MPEG-4 standard was mainly concerned with extraction of video objects, the MPEG-7 standard is concerned with description and representation of these video objects. The MPEG-7 standard defines a generic description language (Description Definition Language, or DDL, for short) for the representation of varied audio-visual objects such as 3-D object representation, scene description, or story structure of a video document. For each different type of video object, there exists different types of description schemes (DS). The MPEG-7 syntax (DDL) supports the multiplicity of these description schemes (DS). To support video object databases (see Figure 1.9), MPEG-7 supports the standardization of representation and search of video objects. Just as the MPEG-1/2 with encoder techniques and the MPEG-4 standard with video object segmentation systems, the MPEG-7 does not advocate any particular form of extraction. Instead, it sup-

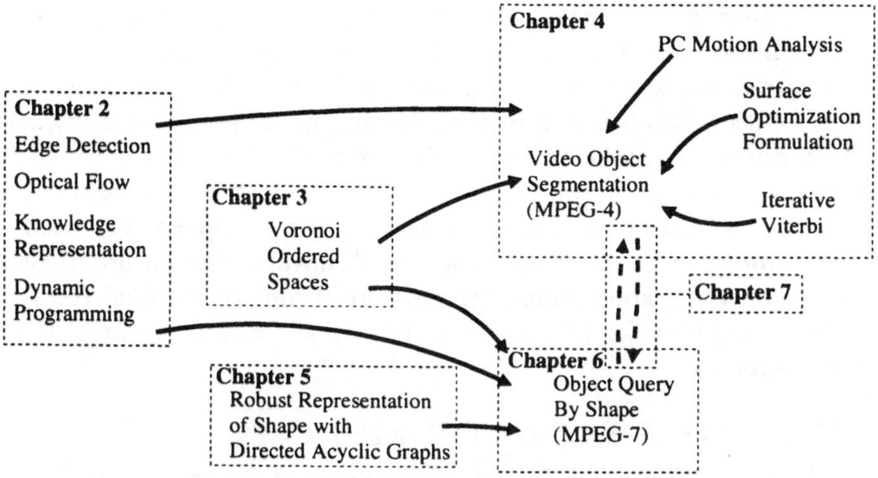

Figure 1.10. Connections between Technologies and their relationship to the key challenges of the MPEG standards.

ports a syntax that meets the demands of a wide range of video object database searches.

This book addresses MPEG-7 issues of representation by implementing robust representation and comparison for shapes of archived video objects and VOPs. In Chapter 3, we present an extraction technique for shape. In Chapter 6, we present a shape query system based upon a multiresolution data representation and its complementary comparison algorithm.

6. SPECIFIC CHALLENGES OF THE MPEG STANDARDS

Of the many challenges of the MPEG standards, the work in this book is centered around two content-based video processing problems: Video Object Segmentation in Chapter 4 and Video Object Query by Shape in Chapter 6, as shown in Fig. 1.10. This book showcases both the theoretical work, Voronoi Ordered Spaces and representation based upon Directed Acyclic Graphs (DAGs), and system design, a Surface Optimization-Based System for video object segmentation and a DAG-based system for Video Object Query by Shape.

VIDEO OBJECT SEGMENTATION FOR MPEG-4

In Chapter 4, we tackle the central challenge of the MPEG-4 standard: video object segmentation. As mentioned before, given a video sequence, a video object segmentation system partitions the space of the video sequence into video objects. In this book, we simplify the general video object extraction to extraction the video objects of different movement.

Our solution involves a formalization of the Video Object Segmentation problem as mathematical surface optimization where the energy function depends on visual information, motion information and object knowledge. Within this system, Voronoi Ordered Spaces are used as means of integrating shape information. A novel variation upon HMM training and the Expectation-Maximization framework called Iterative Viterbi is presented for fitting the surface to the visual artifacts within a video sequence.

QUERY BY SHAPE FOR MPEG-7

In Chapter 6, given a MPEG-7 video object database, we would like to query it by shape, i.e., to ask a user for an outline and search for a VOP that has shape with a similar outline. Shape itself is special case of non-verbal query problem that MPEG-7 framework supports. Our own shape descriptor is one of many descriptors that would likely show up in a interface to MPEG-7 database. The shape query problem has two subproblems that are innately tied together: shape extraction and robust representation/comparison.

The shape extraction is a result from the concept of Voronoi Ordered Space; the shape representation and comparison algorithm were drawn from general methodology inspired by our work in cursive handwriting recognition. Our solution expands the ordered tree paradigm with the introduction of the DAG-ordered tree (or DOT, for short) to construct high-level representations that avoid quantization.

7. RESEARCH CONTRIBUTIONS

In this book, we have made both theoretical and pragmatic contributions to the field of content-based analysis. On the theoretical side, we introduced new concepts in representation, data structure, and shape description. In addition, a number of software systems have been developed to verify the theoretical concepts.

- **Video Object Extraction Techniques**

  1. Voronoi Ordered Space (First Application)
     Based upon the Voronoi cells, the Voronoi Ordered Space is a concept that can specify a descriptive subset of curves that are related to an initial contour. By describing a planar surface in terms of a projection onto a contour and its distance from the

contour, we can order the space w.r.t. an initial contour estimate, enabling our surface optimization algorithm and integrating a concept of shape similarity into our surface optimization.

2. Formulation of Video Object Segmentation as a Surface Optimization Problem
To aid in the analysis of the video object segmentation problem, this book formalizes the video object segmentation problem as a surface optimization problem. This formalization allows for the integration of many different sources (visual, motion, knowledge-based) of video information into a single computational framework.

3. Iterative Viterbi Algorithm
Using the Voronoi Ordered Space of Chapter 3 to describe *a priori* shape constraints, our novel Iterative Viterbi algorithm optimizes a surface by using a dynamic programming algorithm within an iterative framework. For 2-D surfaces that are highly asymmetric in one dimension, our novel Iterative Viterbi algorithm decomposes the surface along the asymmetric dimension (for instance, time in video sequences) into a series of contour optimizations that can be optimally solved, then reassembles the contour estimates into a surface estimate, and recalculates each contour optimization to reflect the continuity of surface.

4. A Novel Video Object Segmentation System
Putting together the formulation of video object segmentation with the Iterative Viterbi algorithm, we produce a system to extract a video object segmentation.

- **Video Object Representation Techniques**

1. Voronoi Ordered Space (Second Application)
By operating upon the Voronoi Order Space, we extract a skeleton-like representation of a shape based upon assumptions of object structure.

2. DAG-Coding and DAG-Compare
Our work in DAGs gives a general methodology for representing the partitioned data [Lin and Kung, 1997a] [Lin and Kung, 1997b] [Lin and Kung, 1998b]. We provide an algorithm for efficient comparisons of these DAG structures. This data structure and its comparison algorithm allow for a robust divide and conquer method and form the basis for our shape query system.

3. DAG-Ordered Trees
From this work in Chapter 5, we also propose a new data structure

called DAG-Ordered Trees. Using the previous work in DAGs in a multi-resolution framework, we can produce a robust shape representation in Chapter 6.

4. A Novel Video Object Query By Shape System
Applying an extraction technique called Voronoi Ordered Skeleton (based upon Voronoi Ordered Spaces), we create a shape query system with novel functionality. Using the robust representation of DAG-Ordered Trees and the extraction of multiresolution representations from Voronoi Order Skeletons of Chapter 3, we implement our video object shape query system.

8. BOOK OUTLINE

The book is outlined as follows. Chapter 1 places the book in reference to various fields of study: Image Segmentation, Computer Vision, Motion Estimation, Neural Networks and Adaptive Signal Processing. Chapter 2 builds a working vocabulary of algorithms and issues associated with video processing. Chapter 3 describes the key concept of Voronoi Ordered Space as a means of both describing *a priori* shape information and extracting shape information from a given contour. Building on Chapter 2 and Chapter 3, Chapter 4 presents a system for Video Object Segmentation. Chapter 5 outlines the general methodology of representing data with Directed Acyclic Graphs (DAGs). Building upon Chapter 3 and 5, Chapter 6 presents a system for Video Object Query by Shape in support of the MPEG-7 standard. Chapter 7 proposes future applications of content-based information processing and links our MPEG-4 and MPEG-7 efforts together, presented in Chapter 4 and 6, respectively, as a single adaptive system.

9. CONCLUSION

We have placed our research within the context of current technologies and directions of future multimedia technologies. Content-based video processing requires a convergence of varied fields of studies to provide the theoretical framework.

A major theme of this book will be the integration of high-level information into low-level computation. From the range of problems given by the MPEG standards, we concentrate upon two major problems, video object extraction and representation that are central to the MPEG-4 and MPEG-7 standards, respectively. From the solutions to these problems, its supporting algorithmic components, and future systems, we hope to shed light how systems translate their high-level representations of knowledge into their low-level extraction operations. Finally, since we are designing media-based technologies rather than vision-based, this work is strongly linked to the current development of MPEG video standards. We continually motivate and measure our work with respect to these standards.

The next chapter overviews the specific technologies that we adopt in this book.

Chapter 2

EXISTING TECHNIQUES OF VISUAL PROCESSING

- Imagine men living in a cave with a long passageway stretching between them and the cave's mouth, where it opens wide to the light. Imagine further that since childhood the cave dwellers have had their legs and necks shackled so as to be confined to the same spot. They are further constrained by blinders that prevent them from turning their heads; they can see only directly in front of them. Next, imagine a light from a fire some distance behind them and a raised path along whose edge there is a low wall like the partition at the front of a puppet stage. The wall conceals the puppeteers while they manipulate their puppets above it.
- So far I can visualize it.
- Imagine, further, men behind the wall varying all sorts of objects along its length and holding them above it. The objects include human and animal images made of stone and wood and all other material. Presumably, those who carry them sometimes speak and are sometimes silent.
- You describe a strange prison and strange prisoners.
- Like ourselves.

Plato's Republic, Book VII

Plato's metaphor of the cave is a literal fact to an engineer involved with video processing. The video sequence is composed of the projections of physical objects. From only one perspective, we must make sense of the world from a series of images as a static three dimensional array of pixel intensities. For such a daunting problem, we are faced with a volume of visual data that only compounds its difficulty. For instance, a single image (a color CIF image, RGB channels of 320x200 resolution) contains millions of points of data; a low resolution color video image at 60 frames per second is on the order of megabytes per second. To make sense of all this data, we must conceptually describe the behavior of the video sequence in a condensed manner via preprocessing. Such preprocessing is further removed from the ground truth than the raw data, but provides insights to the design process. If we accept preprocessing and we are mindful of its biases, preprocessing allows the system and algorithmic designer to understand and better react to the dynamics of content-based video processing. This chapter covers techniques from image segmentation, motion estimation and adaptive signal processing that are the algorithmic building blocks of our software systems.

This chapter is not meant as a complete study of video and image processing techniques, but rather a working language of video and image processing algorithms upon which Chapter 4 and 6 are built. Since we will be working in terms of these algorithms and their output, these forms of preprocessing must have reliable qualities: simplicity, robustness, and graceful degradation. These stringent requirements define a reliable set of algorithms and techniques that we use in our content-based video processing.

1. DESIGN IMPLICATIONS OF HUMAN VISION

An advantage of content-based video processing is that a working system already exists: our own human visual system. Most problems in this book can be easily solved by a well-instructed six-year old child; for a more personal perspective, you can catalog the objects in a room with a quick look. Within the human mind, the content-based processing of object extraction and representation are solved to such a degree that they are taken for granted. However, if we wish computers to process video, we still need to translate our own natural abilities into mathematics, theory, algorithms, and systems.

Our own visual system and pathways guide our research in our content-based video processing. Since the biological visual system is one of the most studied parts of human physiology, we can derive much useful information about how Nature engineered our own vision [Denny, 1994] [Hubel, 1988] [Marr, 1988]. The physical structure of the retina (see Figure 2.1a), and cell differentiations and structures (see Figure 2.1b) demonstrate what cellular specializations are employed for our intake of visual information. Psychophysical results show the internal processing of visual input of the human mind by measuring electrical cell impulses from physical stimuli [Hubel and Wiesel, 1979] [Legge, 1976]. Although the exact mechanisms of the visual pathway are not completely understood, certain dynamics of the human vision can be demonstrated from psychovisual experiments with reaction time and thresholds of recognition [Macleod and Rosenfeld, 1974] [Wilson and Bergen, 1979].

Instead of listing and categorizing all the facts, we will summarize the research work as their relevant implications on our system designs.

How Do We Best Leverage Current Research? Video Object Extraction and Representation must use contributions from many different fields of research.		
Research	Contributions	Caveats
Biological Vision Systems	- A working system - Insights into design process - Insights into visual processing	- Different Implementation (Silicon vs. Cells) - Ambiguous Design Choices
Image Segmentation	- Same problem formulation - Well-tested Feature extraction techniques - Systems that integrate object/scene knowledge	- No temporal aspect -Works with Better Data Quality
Motion Estimation (ME)	- Formalizations of video sequence behavior - alternative representation of video sequence -a prime example of content-based processing	- requires many assumptions - motion info. is ambiguous - Heavy Computation
Knowledge Representation	- another source of information for analysis - a technology for search	- requires classification technologies
Dynamic Programming	- optimization technology - low complexity	- restricted application domain
Compressed Domain	- a source of content analysis w/o computation	- compression and content are not the same
We can use these technologies in our systems, but we must be mindful of their problems.		

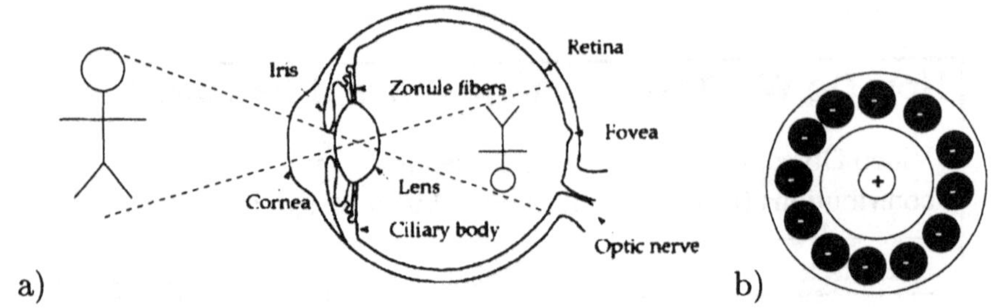

Figure 2.1. Processing Structures of the eye: although we consider the eye, the organ of sight, if we are dealing with the projected form, it is a) the retina that preprocesses the visual information. In b), the plus structures are rods that are sensitive to light regions and minus structures are rods that are sensitive to the dark. Through a specialized excitory/inhibition structure within retina, intensity discontinuity detection can be implemented.

1. **Vision Requires Specialized Hardware**
   Although we think of the human system has ultimately adaptable, the visual system has many functionalities that are fixed at the cellular level. For instance, the retina is an amalgamation of two different types of cells: rods and cones. The rods absorb a broad spectrum of light; the cones absorb a color-specific range. Furthermore, the cones are differentiated by the color absorption of their pigments to sense trichromaticity. Even within the same types of cells, the hard-wired connectivity among cells also varies. Different types of cells have different behaviors toward sustained and transient image patterns. Biological vision is not a single unified process, but rather the management of many sources of information toward the visual objective. When we design our system, a properly managed set of useful primitives may be closer to our own biological solution.

2. **Vision Has Competing Processes**
   Differentiation in vision occurs not only at the cellular level, but also at the process level. Vision itself has two competing processes for bright and dark conditions. In bright conditions, the cones offer us color vision. In dark conditions, the rods, which are more sensitive and have a broader absorption scale, produce most of the visual information and suppress cone activity. Once again, when we design our software systems, we should not be surprised to find competing systems working in parallel.

3. **Vision is Likely Represented in Spatial Coordinates**
   We have naturally defined our video sequence in terms of spatial coordinates. However, we may also interpret the data in terms of the

Fourier coefficients since we achieve good compression rates and good psychovisual quality with frequency-based techniques. The choice of domain greatly affects how we design our systems. Visual testing and biological structure leans heavily toward a representation based on spatial coordinates. Our concept of video space coordinates (x, y, t) match well with how the retina receives the visual information.

4. **Vision Supports the Differential as a Basic Operation**
   The human visual system has an amazing visual operating range that spans over 10-11 logarithmic scales of light intensity. The key to the broad operating range is that the retinal structure implements a local differential operation by placing the two different types of cells within an excitory-inhibitory structure (see Figure 2.1b). For instance, a green-red difference is implemented by placing a green cone within an excitory region surrounded by red cones in an inhibitory region. The spatial difference operation is clearly a basic visual operation of the eye.

5. **Vision is a Temporally Linked Process**
   The retina has its own transient state. Although we can consider the video sequence as series of images, an isolated image does not have an equivalent effect of an image within the context of the sequence. This effect of previously seen images on the retina can last beyond 100 milliseconds. Time is not only a construct of the mind, but also a physical attribute of visual acquisition devices in our retina. These effects show the importance and scale of temporal correlation that exists between sequential frames.

In this section, we have outlined some surprising features of our own visual system. We considered only low-level processing because 1) we wished to avoid misinterpretation of the biological results and 2) the bulk of the research work and understanding of vision is concentrated at the cellular level. Although few insights into the high-level processes of human vision are given, we can understand some of biological visual preprocessing.

2. IMAGE SEGMENTATION

Image segmentation is the precursor to content-based information processing. Image segmentation is the extraction of VOPs from a single frame of video. By extending these image segmentation techniques to handle multiple frames, we can begin to design our content-based video processing systems. Furthermore, image segmentation has many common issues as our content-based video processing: duality in problem

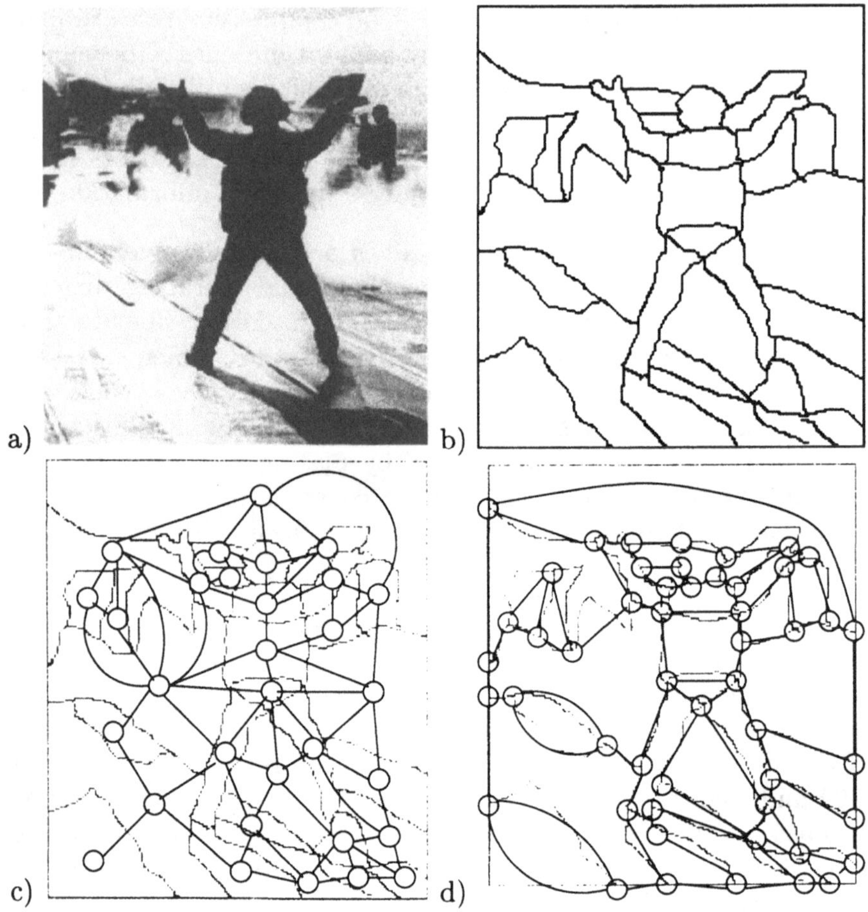

Figure 2.2. Different Graphical Interpretation of the Image Segmentation Problem: a) the original image b) the first phase analysis, using spatial correlation of intensity to create regions or identify region boundaries c) the graph created by a region-based representation: an object is represented by a set of connected components and d) the graph created by a boundary-based representation: an object is represented by a closed cycle

statements and its connections to graph theory, the extraction of image features through intensity and spatial correlation, and the integration of object knowledge for image understanding [Fu and Mui, 1980].

DUAL PROBLEM REPRESENTATION

There are two equivalent ways of stating the image segmentation problem. The choice of problem statement can dictate the applicable techniques and algorithms. We can map the problem of image segmentation in Fig. 2.2 onto two different graphs and note the different oper-

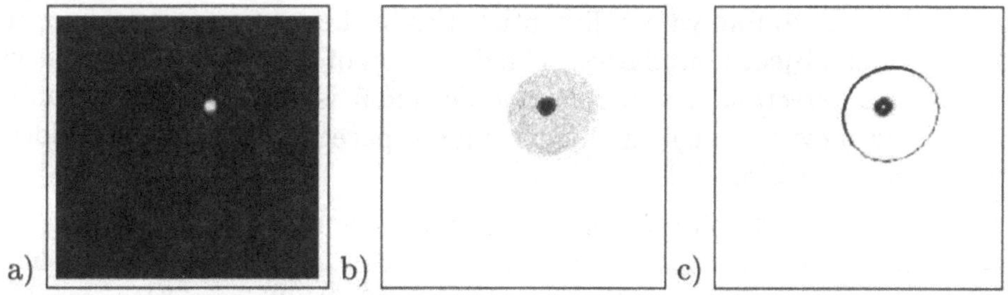

Figure 2.3. Limits of Spatial Correlation of Intensity: a) the original image of a lit sphere. Although the sphere is a single object, because of the lighting, b) the clustering of image intensities and c) edge detection values suggest multiple objects. Identification of object boundaries is clearly not all visual.

ations that we use. In Fig. 2.2c, the *region-based* representation maps the regions to nodes and the boundaries between regions to edges; in Fig. 2.2d, *boundary-based* representation maps boundaries to edges and boundary junctions to nodes. The form of solution depends upon the graphical mapping. For region-based problems, the solution becomes a matter of finding connected components, involving algorithms such as clique finding, independent set and graph coloring algorithms. For a boundary-based representation, the operations are a matter of a finding closed path, such as Depth First Search, and minimum cut [Cormen et al., 1990]. Regardless of formulation, both problems are computationally difficult, since path-finding such as Hamiltonian Path and clustering problems such as clique finding are both NP-Complete [Garey and Johnson, 1979]. However, the algorithmic design space is determined by the choice of region-based or boundary-based representation. In our own work, the video object segmentation problem has the same duality and the same issues, mentioned above, also apply.

USING SPATIAL INTENSITY PATTERNS: CLUSTERING/ EDGE DETECTION

The two processes, *clustering* and *edge detection*, are two techniques that derive object features from spatial intensity patterns. Spatial correlation of intensity is our most precise and robust feature in our video sequence, strongly supported in image processing and in the biology of our own human visual system. Like the problem representation, this type of image processing also has a dual form. Using the correlation of intensity and locality to infer to which video object the pixel belongs, we can locally *cluster* pixels into regions that belong to the same video object. Assuming the pixel *clustering* is correct, we need only to determine the video object membership of these regions. The dual of the clustering problem is to find where discontinuities in the spatial patterns occur due to video object boundaries. The detection of these discontinuities is called *edge detection.* Assuming *edge detection* is correct, we need only to consider how the edges divide the video space to determine the video object membership.

Predicated on the simple assumption that regions of the same intensity are derived from the same object, clustering can identify subregions that belong to the same object [Duda and Hart, 1973]. This problem can be a multidimensional clustering problem that can include intensity, color, and position. However, there are limits to clustering on the basis of image intensities. Clustering assumes that the pixels from the same video object are independent w.r.t. their intensity and position within the object. Certain intensity patterns may break that assumption: a simple reflective surface of a sphere (see Figure 2.3b) or a repeating pattern. In those cases, oversegmentation may result and the concept of clustering must give way to higher levels of modeling and processing.

Edge detection is a common operation in image processing. The concept of an edge is overused and the definition of an edge varies from field to field. In some circumstances, edges are meant to be exact as with the concept of zero crossings[Marr, 1988]. The concept of edge detection that we use in this book is probabilistic: if $E(x, y)$ is an edge detected image, then the value of $E(x, y)$ is proportional to the probability that a boundary between video objects exists at (x,y) for that given time. Thus, our specific application of edge detection is to find inter-object boundaries. Variations upon the partial differentials applied to $I(x, y, t)$ in the x and y dimensions are most useful in our applications. Like clustering, edge detection assumes that the pixels from the same video object are independent w.r.t. their intensity and position within the object, and suffer from the same problems as clustering (see Figure 2.3c). One of the advantages of edge detection is that only the discontinuity (a 1-D structure for images) needs to be modeled rather than intensity behavior over a two dimensional region.

OBJECT KNOWLEDGE

Since edge detection and clustering are localized analysis, image segmentation assembles and rejects localized analysis through global constructs such as connectivity, area thresholding, and high-level knowledge of the object in question. Image intensity patterns may shift beyond a concept of local variance and yet remain a part of the same object; edges that are oriented to form a closed contour are likely to be part of an object boundary. We can strengthen our claims of correlation (or dis-

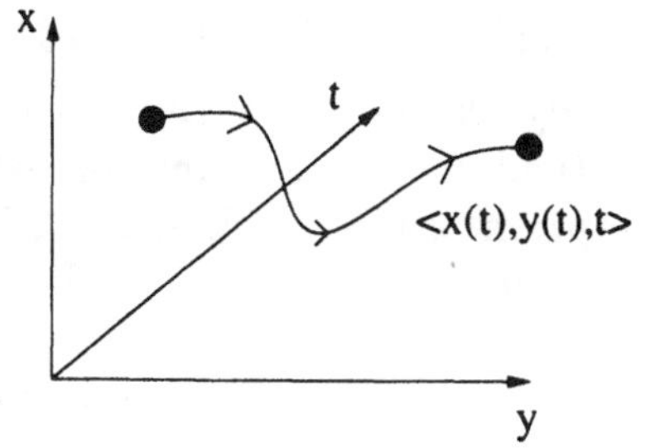

Figure 2.4. The optical flow constraints assumes that a point that moves along a path in 3-D space of video sequence has a constant intensity.

continuity) by either modeling the correlation of intensity and position (or integrating the known structure of the object into analysis).

First, we consider concepts such as locality and connectivity that do not require *a priori* knowledge of the object. These types of analyses enhance or reject the local information through the consideration of a larger context. For a region-based approach, if regions share a large proportion of their border, then differences in their intensities may be overcome and these regions may be joined together. For an edge-based representation, if we can find a set of edges that form a closed contour, then these edges are more likely to belong to an object boundary. Such analysis includes region growing algorithm [Zucker, 1976], snake algorithms [Kass et al., 1988], edge following algorithms, pyramidal image representation [Pappas, 1992] [Bouman and Liu, 1991] [Horowitz and Pavlidis, 1976] and model fitting [Hotter and Thoma, 1988].

For the knowledge-based techniques that use *a priori* information, we defer our discussion to section 4., since this type of analysis intertwines both segmentation and representation.

3. MOTION ESTIMATION

Unlike images, video has the concept of motion and we can use it as a strong discriminant of object membership in the video space. In some cases, motion alone can separate a video sequence into video objects. In the previous section, we studied techniques to exploit spatial locality; in this section, we now exploit the spatio-temporal correlation. Using the assumption of the existence of the physical object, we can trace the movement of a physical object through the video space as a set of separate parallel paths under certain conditions.

THE ASSUMPTION OF OPTICAL FLOW

The movement of a physical object through time can be embodied in the assumption of optical flow. Many motion estimation techniques use this implicit assumption within their calculations. This assumption is described mathematically by the optical flow equation [Horn and Schunck, 1981], shown in Eq. 2.1. Given a pixel, $\langle x(0), y(0), 0\rangle$, we can associate a path through time that describes its projected motion:

$$I(x(t), y(t), t) = c \tag{2.1}$$

where c is a constant and $\vec{f} = \langle x(t), y(t), t\rangle$ is the path of the pixel parametrized with respect to time (see Figure 2.4). Eq. 2.1 also has another form from the Taylor's series expansion and the application of the chain rule:

$$\frac{\partial I}{\partial x}\frac{dx}{dt} + \frac{\partial I}{\partial y}\frac{dy}{dt} + \frac{\partial I}{\partial t} = 0 \tag{2.2}$$

where second order terms are considered negligible. Eq. 2.2 relates the projected motion of the video object, $\left\langle \frac{dx}{dt}, \frac{dy}{dt} \right\rangle$ to the partial derivatives of $I(x, y, t)$. In this book, when we mention object motion, we refer to the projected object motion onto the frame, $\left\langle \frac{dx}{dt}, \frac{dy}{dt} \right\rangle$.

The optical flow assumption also leads to an alternate representation of the video sequence where pixels are grouped with respect to the projected motion. Assuming the motion between frames is correct, we can group sets of pixels by their time parametrized paths of constant intensity. For a video sequence, we can group these pixels by their paths that best satisfy the optical flow constraints; these paths are called *motion paths*. This grouping of pixels by motion path is a convenient form of the video sequence.

The issues with motion estimation algorithms based upon the optical flow are:

1. when using Eq. 2.2 to find the projected motion, the problem becomes a minimization problem with two unknowns and only one constraint.

2. the optical flow equation assumes that the intensity of a point is constant over time. But there are many counterexamples: a rotating reflective surface, changing lighting conditions, shading, etc.

When we use optical flow in the system, we need to address the issues mentioned above. We usually add another constraint, such as motion

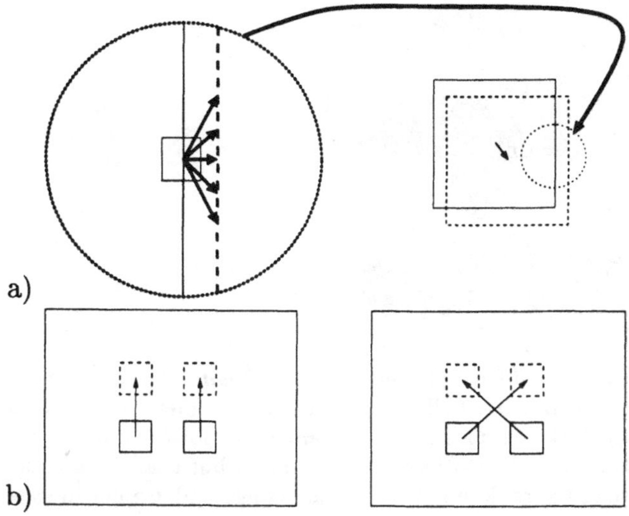

Figure 2.5. Two fundamental ambiguities of the optical flow constraint are demonstrated where position of the objects the next frame is shown by the dashed boxes. a) The Aperture Problem: If we try to determine the movement of the middle of a side, the correspondence of a block in the next frame is ambiguous. However, if we zoom out and take a larger context, the movement can be clearly resolved. b) Correspondence Problem: The movement of two identical blocks upward has two explanations: either both blocks moved upward as on the left, or they diagonally switched places between frames as on the right.

smoothness, to Eq. 2.1. The effect of changing lighting conditions cannot be avoided and these lighting conditions must be caught before optical flow processing. In our work, most of our test sequences do not have such difficult lighting conditions.

However, even with these assumptions, there exists two inherent problems with the optical flow assumption: the *aperture* problem and the *correspondence* problem (to be defined, shortly). These problems are related to the fact that although projected motion is considered a low-level feature, motion contains much high-level information such as object membership and object motion.

As shown in Figure 2.5a, the *aperture* problem is caused by the lack of the contextual information to resolve visual ambiguities. For example, consider a section of the homogeneously-filled square moving from one image to the next. If we consider a macroblock from only one side of the square in isolation, its motion is ambiguous. Without knowledge of the corners, it is difficult to predict the motion of the sides. Without knowledge of the sides, it is difficult to predict the bulk of the square. Both are cases of the aperture problem. Note that if there was a vertical

a) 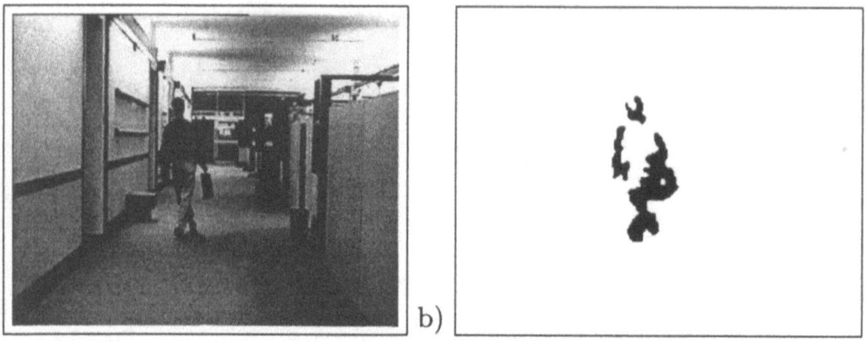 b)

Figure 2.6. Example of Change Detection: a) Original Image of the video sequence, a person walking through a hall with a stationary camera; b) the Change Detection image where the darker regions indicate changing regions. Note that the human figure is clearly located from the stationary background, but that the homogeneous regions within the human figure do not have visual change and are not marked as changing.

grayscale gradient in Fig. 2.5a instead of a homogeneous fill, then there would be no aperature problem on the inset.

As shown in Figure 2.5b, the *correspondence* problem is an inherent problem of ambiguity that cannot be resolved through visual context. Consider two identical gray squares that move through the frame. In the next frame, there are two gray squares that have been relatively displaced up. Our natural inclination is that both squares have moved up. However, the optical flow constraint allows for two possibilities, i.e., our natural inclination and the possibility that the two squares have changed places between frames. Although this case seems far-fetched on a large scale, any two macroblocks with similar features are allowed to do this type of switching, according to the optical flow equation.

To resolve these ambiguities, we need a secondary constraint. These issues will be addressed within the implementation of motion estimation algorithms in Section 2.7.

CHANGE DETECTION

Change detection is a simple, but powerful type of motion analysis, as shown in Figure 2.6. If our background is stationary, i.e., $\frac{dx}{dt} = 0$ and $\frac{dy}{dt} = 0$, motion paths associated with the background are parallel to the t axis and follow a simplified form of the optical flow equation:

$$\frac{\partial I}{\partial t} = 0 \tag{2.3}$$

Therefore, non-background pixels have a non-zero values of $\frac{\partial I}{\partial t}$. There are many schemes to detect change through the behavior of $\frac{\partial I}{\partial t}$ as a

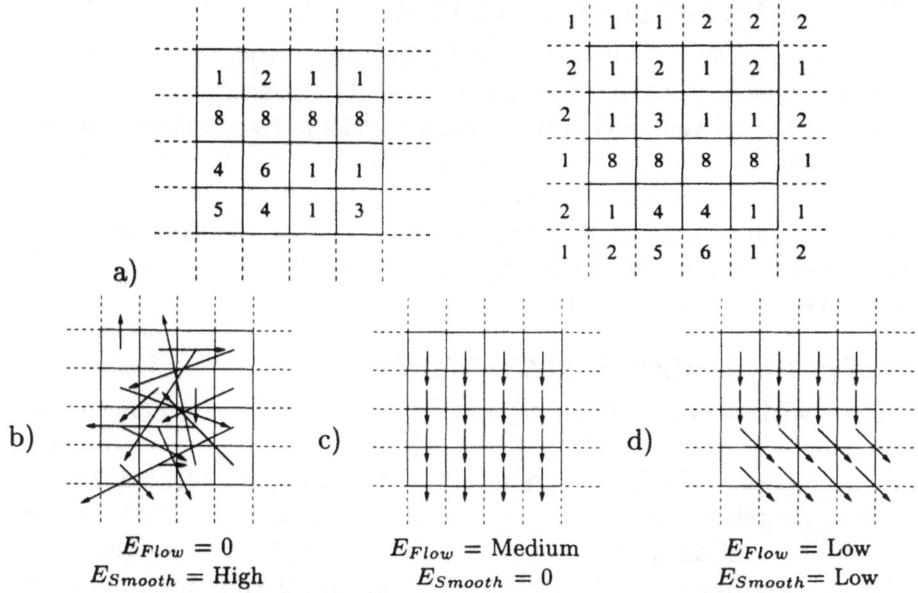

Figure 2.7. Balancing smoothness and optical flow for Motion Estimation. Given a) the intensities of a section of a frame and its following frame, we can rate a given motion field with respect to Eq. 2.5 and Eq. 2.6. In b), we have a zero error with respect to the optical flow, while creating an incoherent motion field. In c), as in block matching, we have completely smooth motion field and incur optical flow cost. In d), a balance between smoothness and optical flow result in the most accurate motion field.

function of time. The simplest of these schemes is as follows: watch a given spatial location on the frame, mark that pixel location as non-background if the intensity of pixel changes beyond a threshold. For non-stationary backgrounds, we can compensate the background motion through offsetting of the frames, and still use change detection to find non-background regions. However, the reliance on motion compensation compromises the robustness of change detection since the change detection now depends on the accuracy of motion compensation. Some movement is not equivalent to change: for instance, homogeneous regions may not exhibit any change properties through temporal correlation. Also, change detection is a binary operator that distinguishes between only two different classes (background and non-background) of objects and may misgroup objects in scenes with three or more object classes. An example of a system that uses signal analysis upon $\frac{\partial I}{\partial t}$ to separate background and foreground is a scheme based upon a high ordered statistics (HOS) [Neri et al., 1998].

TWO METHODS OF MOTION ESTIMATION

The calculation of the *motion field*, i.e., the projected motion over the video space, must balance the optical flow constraint with *a priori* knowledge of the motion field characteristics. As mentioned in section 2.4, we cannot determine two variables (x and y components of projected motion) with only one constraint. Two methods of motion estimation, block matching and optical flow balance the optimization of the optical flow constraint with a smoothness criterion (see Figure 2.7) to calculate the motion field.

METHOD 1: BLOCK MATCHING

Along with the optical flow constraint, *block matching* assumes that a given neighborhood (usually, a n-by-n block) around a point moves with the same motion. Under the assumption of a constant projected motion over the neighborhood, we can determine the motion of a given neighborhood by finding the Least Square Fit to the optical flow equation.

Block matching computation is as follows. By dividing the image region into a regular grid of n-by-n pixels, one can use a block-matching algorithm to find the motion field for a given frame. Let us consider the two frames of the video sequence, $I_{current}(x,y) = I(x,y,t)|_{t=t_o}$; $I_{next}(x,y) = I(x,y,t)|_{t=t_o+\Delta t}$. For simplicity, consider a neighborhood N that is centered around the origin (for a n-by-n block, $N = \{-\frac{n}{2},\ldots,\frac{n}{2}\} \times \{-\frac{n}{2},\ldots,\frac{n}{2}\}$). From Eq. 2.1, we quantify the error of a given motion vector by the intensity difference of the neighborhood (N) of the current frame and an offset neighborhood in the next frame.

$$SSD(\vec{\delta}) = \sum_{\vec{n}\in N} \left(I_{next}(\vec{n}+\vec{\delta}) - I_{current}(\vec{n})\right)^2 \tag{2.4}$$

where SSD is short for the Sum of Squared Difference, $\vec{n}$ is the 2-D offset vector of each pixel in the neighborhood, and $\vec{\delta}$ is the 2-D motion vector that is constant over the neighborhood. To find the motion vector by block matching, we wish to find $\arg_{\vec{\delta}} \min\left(SSD(\vec{\delta})\right)$. To compute the whole field, we repeat the calculation with the proper offset for each block in the motion field. Assuming some upper limit on the magnitude of the motion field velocity, the SSD minimization is generally computed through an exhaustive search, although more efficient schemes do exist [Chen et al., 1991] [Jain and Jain, 1981].

Problems with block matching are due to the assumption of constant velocity over the neighborhood. In addition to the optical flow assumptions, the locality assumptions about the neighborhood can be violated as well because:

1. the neighborhood contains two or more differently moving objects.

2. the ground truth motion is not smooth. Consider an extreme case of the center of a spinning wheel; motions within that area are clearly not equal.

3. the neighborhood is too small and SSD has multiple minimal solutions Consider the block inside a homogeneous region.

For block matching, there exists a trade-off in neighborhood size. Issue #3 can be solved by increasing the neighborhood size. However, a larger neighborhood size also increases the probability that issues #1 and #2 are problematic. Solutions to this trade-off involve the integration of high-level knowledge and is beyond the scope of block matching algorithms.

METHOD 2: HORN AND SCHUNCK ITERATIVE MOTION FIELD OPTIMIZATION

To find the motion field via calculus of variations, we formulate the calculation of the motion field of a given frame as an optimization problem.

The formulation in this section is used in the original work by Horn and Schunck [Horn and Schunck, 1981]. We use the optical flow constraint as one optimization criterion and we translate a secondary constraint of smoothness as another. The linear combination of the two criteria yield the function to be optimized. The error for the optical flow constraint can be quantified from Eq. 2.2:

$$E_{flow}(x,y) = \left(\begin{array}{c} u(x,y) \cdot \frac{\partial I}{\partial x}(x,y,t)\Big|_{t=t_\circ} + v(x,y) \cdot \frac{\partial I}{\partial y}(x,y,t)\Big|_{t=t_\circ} \\ + \frac{\partial I}{\partial t}(x,y,t)\Big|_{t=t_\circ} \end{array} \right) \tag{2.5}$$

where $u(x,y)$ and $v(x,y)$ are the x and y components of motion field over a single frame at time $t_\circ$, respectively. E_{flow} underconstrains the problem; we add smoothness component as the magnitude of the gradients of u and v.

$$E_{smooth}(x,y) = \left(\frac{\partial u}{\partial x}\right)^2 + \left(\frac{\partial u}{\partial y}\right)^2 + \left(\frac{\partial v}{\partial x}\right)^2 + \left(\frac{\partial v}{\partial y}\right)^2 \tag{2.6}$$

Now that we have a measure of both the smoothness and optical flow error, we can linearly combine the two metrics over the motion field of a given frame.

$$E_{total}^2\left(u(x,y),v(x,y)\right)=\int\!\!\int\left(\alpha^2E_{smooth}^2+E_{flow}^2\right)\,dx\,dy \tag{2.7}$$

where α is a weight dependent upon the variance of intensities over the video sequence.

Our motion field calculation can be described as follows. For a given frame, we wish to find:

$$\arg\left(\min_{u(x,y),v(x,y)} E_{total}^2(u,v)\right) \tag{2.8}$$

To find the motion field that minimizes this function E_{total}, we apply the fundamental theorem of calculus of variations and the iterative solution of the Gauss-Seidel. If we approximate the second gradient as $\nabla^2 v = v-\bar{v};\nabla^2 u = u-\bar{u}$, where $\bar{v}$ and $\bar{u}$ are the averaged x and y motion components, then we can iteratively optimize our energy function with the updating scheme:

$$u^{(n+1)}=\bar{u}^n-\frac{I_x\left[I_x\bar{u}^n+I_y\bar{v}^n+I_t\right]}{\alpha^2+I_x^2+I_y^2} \tag{2.9}$$

$$v^{(n+1)}=\bar{v}^n-\frac{I_x\left[I_x\bar{v}^n+I_y\bar{u}^n+I_t\right]}{\alpha^2+I_x^2+I_y^2} \tag{2.10}$$

Optical flow calculations can give a dense and an accurate motion field. Furthermore, unlike block matching, this method propagates the local smoothness information through the whole frame iteratively, resolving the aperture problem in some areas. The information propagation comes at a cost: the Horn and Schunck optical flow calculations are computationally intensive and the computation is iterative, disallowing speed-up by parallelism.

Although Horn and Schunck optical flow motion field is one of the more accurate and dependable extraction techniques, there are certain places where it is inherently wrong. Although our secondary constraint is more flexible than the block matching assumption, there are places in the motion field where the smoothness constraint is incorrect. For instance, at boundaries of two differently moving and overlapping objects, the motion discontinuity directly conflicts with our smoothness criterion. To attenuate this problem and improve motion field accuracy, there are many more advanced methods available [Nagel, 1987] [Ong and Spann,

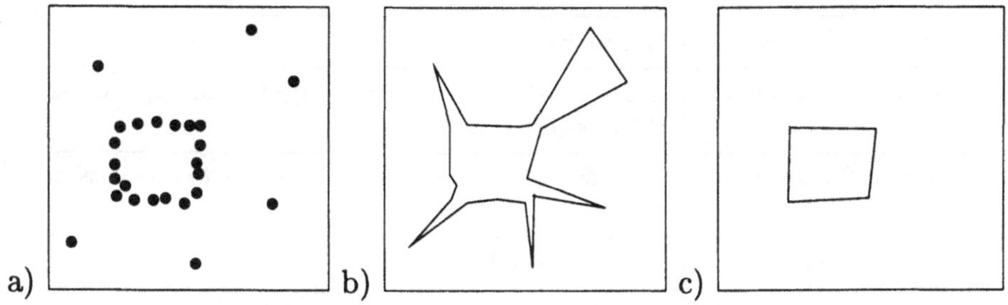

Figure 2.8. Conceptual Role of Knowledge: a) some points that may be part of an object boundary or noise, b) if we assume the object boundary goes all points, we are fitting our boundary to noise. c) if we knew the boundary is smooth, we can reject noise and find a better boundary estimate.

1996] [Barron et al., 1994] [Aggarwal and Nandhakumar, 1988] [Lai and Vemuri, 1995]. In our own work, we account for these inaccuracies and use supplemental motion estimates to avoid this problem.

4. KNOWLEDGE REPRESENTATION

While visual artifacts are important in content-based video processing, *a priori* object knowledge can be used to refine visual artifacts and separate content from noise (see Figure 2.8). For the last two sections, we have focused upon using spatial and spatio-temporal patterns of the video sequence to infer object features. Our third class of analysis is based upon the unseen, i.e., not directly derived from $I(x, y, t)$. In this chapter, we extend our analysis of the video sequence by representing *a priori* knowledge of the object and leveraging these representations into our analysis. Either through assumptions about the qualities of video objects, generalizations of structure over classes of objects, or *a priori* knowledge of the object class, the motion and visual information can be assembled together in an intelligent manner. This section covers a number of object representations/models that are available to the designer and the key issues of how to use these models for content-based analysis.

AVAILABLE REPRESENTATIONS

Representations such as parametric models and neural structures can explain spatial and temporal patterns within the video sequence as artifacts of object structure. What is known can sometimes be more helpful than what is seen. For instance, when watching analog television, noise is sometimes seen, but we know to dismiss such visual artifacts on our television as "bad reception." We can so easily reject the noise because television content fits an underlying model in our mind. As shown in Fig-

Type of Model	Object Requirements
Spatial Smoothness	Disallows sharp and jagged objects and approximates them with a smooth outline.
Temporal Smoothness	Disallows sharp and sudden motion movements and approximates them with a smooth motion.
Region Thresholding	Disallows small objects.
Affine Motion Models	Allows only 3-D projection of rigid objects.
Kalman Filter	Tracks motion objects that moved with a second-order input (acceleration or force in a physical system) with a good initial estimates. Applies to all moving unoccluded physical objects.
Human Walking Model	Accounts for human subject bending, movement, and gait. However, human subject **must** be walking orthogonal to camera projection.
Voronoi Ordered Spaces	Requires a good initial estimate of the object shape

Figure 2.9. Models and their respective classification requirements: Knowledge models are only as useful as when they are applied to the correct situation. For every model, one must also consider the cost of classification in its utility.

ure 2.9, these representations range from simple physical laws to specific object-based mechanisms and connectivity. Connectivity and region-based thresholds allow for the global-based analysis of boundaries and regions to supersede the local analysis. Implications of physical laws can be mathematically integrated into analysis. For example, Newton's first law of motion can be expressed in terms of a second-order Kalman filter system. For rigid objects, an affine motion model can account for variations in the motion field and provide a fitness criterion for object membership. If we know a human being figure is walking orthogonal to the camera view, then we can leverage our knowledge of human structure into our extraction techniques [Cheng and Moura, 1999].

However, these models are only useful when applied to suitable situations. When applied incorrectly, the object representation may degrade system performance. Smoothness constraints may remove sharp protrusions that belong to an object. If the object becomes occluded, Kalman filtering may track a ghost image. An affine model may split a bending object into two separate objects. Expressing or coding the models themselves is usually not the problematic step in using them. The utility of the model/representation is strongly related by our ability to classify, i.e., to find the correct model for the given content.

CLASSIFICATION

When using *a priori* knowledge of an object for representation or extraction, we require classification technologies. If we choose the correct object representation, then *a priori* knowledge from an object representation can assemble partial information derived from the video sequence with better noise rejection and tolerance to missing visual information. If the object class is not known, incorrect application of a representation actually degrades analysis by disallowing otherwise valid video object features or analysis. For instance, the fitting of a human walker model to the video sequence can locate all body parts of the walker and even his gait, but only when the person is walking perpendicular to the camera [Cheng and Moura, 1999]. How to determine whether the applicability of a model and the reliability of results are still an open question. In contrast, temporal smoothness applies to most objects and, if it is not completely correct, degrades system performance in a graceful manner.

The issue of object classification can be avoided either 1) by the generality of the model or 2) by restricting the input to a certain object class. These two options either limit the expressiveness of the model by enforcing generality as a constraint on the model or fail to balance the specificity of a model over a wide range of data, respectively. Simple generalizations about the projected object behavior may mislead in a small percentage of cases, but provide essential error resilience and rejection in the rest. By restricting the input class, the presupposition of object class only reinforces the importance of a working classifier.

Classification technologies are not only important tool for extraction and recognition, but also an important query functionality for the "new" media of the future. In extraction, classification of unknown objects allows the selection of the proper object-specific model that, in turn, leverages object-specific *a priori* knowledge into our analysis. In representation, such a classification technique can also be used as a query tool. In Chapter 3, we present the concept of Voronoi Ordered Spaces that fulfills the dual purpose of extraction and representation for both our video object extraction and query.

5. DYNAMIC PROGRAMMING

The dynamic programming technique for optimization is used both in Chapter 4 for VOP extraction and Chapter 6 for shape query. In 1955, R. Bellman introduced the concept of dynamic programming for optimization of certain problems [Bellman and Dreyfus, 1962] [Amini et al., 1990]. It is a powerful and computationally efficient method for optimization that can not only find the optimal solution for the optimization problem, but also solves all the optimal subproblems and returns their relationship to the final optimal solution.

The dynamic programming technique can only be applied only if two particular qualities of the optimization problem are satisfied: *optimal substructure* and *overlapping subproblems* [Cormen et al., 1990]. The *optimal substructure* property requires that "an optimal solution to the problem contains within it optimal solutions to subproblems [Cormen et al., 1990] (p. 309)." By solving optimal subproblems, dynamic programming can calculate the optimal solution. The *overlapping subproblem* property assumes that there are finite number of subproblems that can be produced and their computation can be shared among all the optimal subproblem calculations. This reuse of computation in subproblem dependency and subproblem definition allows the dynamic programming to be efficient and optimal. However, for optimization problems that do not have these properties, we must first translate the problem into dynamic programming form. Although we gain optimality from the dynamic programming technique, the translation often affects the quality of the final solution.

6. COMPRESSED DOMAIN

Although there exists a strong correlation between coding and representation, it is important to realize where coding for compression and content-based information processing converge and diverge. In section 5., we noted how the MPEG-1/2 syntaxes were related to our query and extraction efforts. Since the majority of video data will be in the compressed form, we wish to operate in the compressed domain. However, we must be wary of those coding efforts (such as MPEG-1/2) whose focus is on coding efficiency.

The compressed domain representation of a video sequence often contains preprocessed information that can aid content-based processing. However, when working in the compressed domain, we must assume that we cannot reference the original video and must be wary of mistaking coding artifacts for content. On the encoder side, coding artifacts may be embedded into the data, especially when the coding effort is aimed toward coding efficiency and not toward content-based functionalities. In cases where the compressed domain features are straightforward representations of the video content such as DC coefficients of the DCT components macroblocks, analysis upon the compressed domain can leverage the encoder processing into content-based video processing [Yeo and Liu, 1995]. Other compressed domain features (macroblock references) features may not correspond to the ideal content information (the projected motion of an object). For instance, in MPEG-1/2, although macroblock references in P and B frames correlate well with object motion, we should recognize that these macroblock references are calculated for greatest coding efficiency. Although much motion information can be derived from the macroblock references, macroblock references are not equivalent to motion. As the compressed domain syntaxes (such as the MPEG-4 and MPEG-7 standards) move toward content-based functionalities, the aims of compression and content will converge, and such issues become much less problematic.

7. CONCLUSION

In this chapter, we have discussed techniques and issues to derive visual information and integrate *a priori* knowledge into our content-based analysis. This chapter establishes a set of video processing techniques that facilitates our discussion of our system designs. The chapter also confronts the issues that are associated with preprocessing and the implications upon our system designs. By the inherent nature of content information processing, we are moving away from the ground truth of the video sequence with this preprocessing. However, as long as we are mindful of the issues with this preprocessing of our video sequence, we can use the techniques of this chapter to simplify analysis and our system designs.

In the next chapter, we discuss the theoretical concept of Voronoi Ordered Spaces. Combining these existing techniques within our contributions of Voronoi Ordered Spaces in Chapter 3 and DAG Representations in Chapter 5, we create novel systems for content-based processing: a video object segmentation system in Chapter 4 and a shape query system in Chapter 6.

Chapter 3

VORONOI ORDERED SPACE

Starting from principles of Euclidean geometry, this chapter introduces a central concept of Voronoi Ordered Space in our book. The Voronoi Ordered Space is a $\Re^2$ space that describes a point in $\Re^2$ in terms of its projection onto a contour-based description of a shape. With the amount of work and analysis done in the area of shape description, this chapter is a subset of previous work (listed in Section 1.). We use the concept of Voronoi Ordered Space as our interfaces for shape information in our system designs: to integrate shape information into low-level optimization algorithms in Chapter 4 and to derive shape representations from video object planes in Chapter 6. Our contribution is the demonstration of how Voronoi Ordered Space can express concepts of shape through concrete system functionalities.

1. PREVIOUS WORK

The relationship between Euclidean distance and skeletonization has been studied with the Medial Axis Transform (MAT) (also known as the grassfire algorithm) [Blum and Nagel, 1978] [Blum, 1973] [Malladi et al., 1995] [Lee, 1982]. The hierarchical representation of MAT-derived shape was studied [Shapiro, 1980]; this work culminated in an excellent treatment that clearly shows its multiresolution properties [Ogniewicz and Kubler, 1995]. Related concepts such as Voronoi Cells and thinning processes are well-known in image processing [Aurenhammer, 1991] [Rosenfeld, 1986].

Our contribution is the demonstration of how Voronoi Ordered Space can express concepts of shape through these concrete system functionalities. This chapter simplifies and condenses a subset of the previous analysis, packaging it into a single concept called Voronoi Ordered Spaces.

While ignoring some of the deeper mathematics, we limit our analysis to the design of our two systems for video object extraction and representation. Using analysis from this chapter, Chapters 4 and 6 show how the Voronoi Ordered Spaces can express shape information within system design.

2. NOTATION

First, we will be working in a $\Re^2$ space that represents an image. Notation for this chapter is as follows:

1. Lowercase English letters $(a, b, \ldots)$ are scalars.
2. Lowercase Greek letters $(\alpha, \beta, \ldots)$ are functions that return scalars.
3. Lowercase English vectors $(\vec{a}, \vec{b}, \ldots)$ are points in $\Re^2$.
4. Lowercase Greek vectors $(\vec{\alpha}, \vec{\beta}, \ldots)$ are functions that return points or vectors, depending on the context.
5. A pair of uppercase letters are a line segment between the points of corresponding lowercase letters, e.g., AB is the line segment between points $\vec{a}$ and $\vec{b}$.
6. A single uppercase letter $(A, B, \ldots)$ is a set of points such as contour or an area.
7. A boldface single uppercase letter $(\mathbf{A}, \mathbf{B}, \ldots)$ is a set of sets of points such as a set of contours.
8. Uppercase Greek letters $(\mathbf{\Lambda}, \mathbf{\Phi}, \ldots)$ are functions that return a set of sets of points such as sets of areas.

3. DEFINITION OF VORONOI ORDERED SPACE

We define a restricted set of contours to represent our shape.

DEFINITION 3.1 (THE CONTOURS OF $\mathbf{C}$) *We define* $\mathbf{C}$ *as the set of contours that are defined below:*

1. *a simple closed contour in* $\Re^2$
2. *piecewise continuous.*
3. *parametrized by arc length, such that* $C = \{\vec{\gamma}(s) | 0 \leq s < \|C\|\}$, *where* $\|C\|$ *is the length of the contour* C.

How do we use Voronoi Order?

Voronoi Order is a means of accessing a 2-D space, much like Cartesian Coordinates.

A point Z in the 2-D space can be uniquely identified by its projection onto an onto an arbitrary line, the x-axis.

The X coordinate is the distance between the origin and the projection on the x-axis. The Y coordinate is the distance from the x-axis.

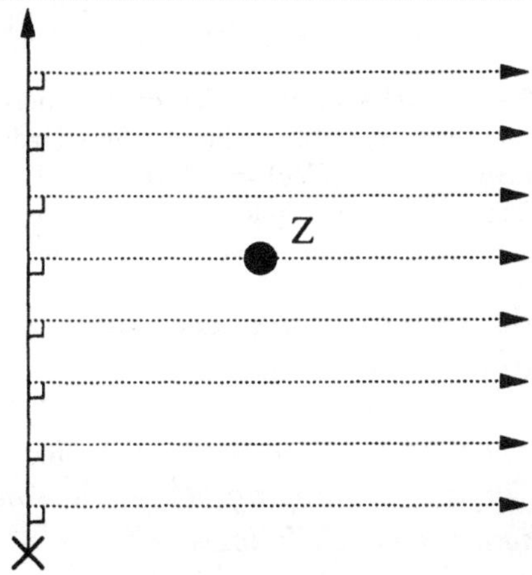

We can use an arbibrary contour C for the x-axis , instead of a straight line. For a point Z, the X coordinate in this alternative coordinate system is called the Voronoi Order; the Y-coordinate, the Voronoi Distance.

In our video object extraction system, we apply our calculations on this new coordinate system and integrate a concept of shape into our analysis.

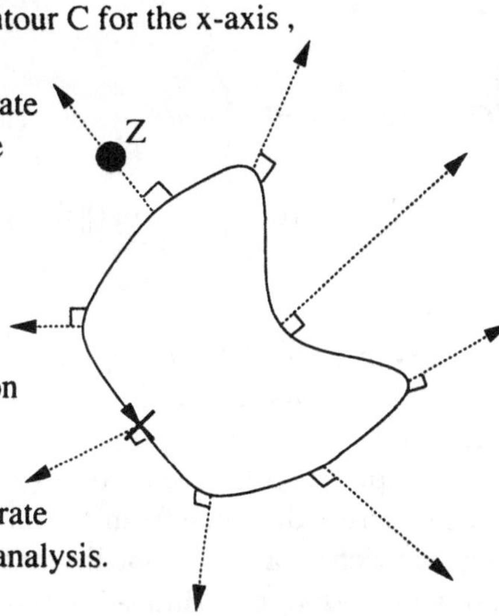

Voronoi Order allows us to seamlessly integrate the concept of shape into our computations.

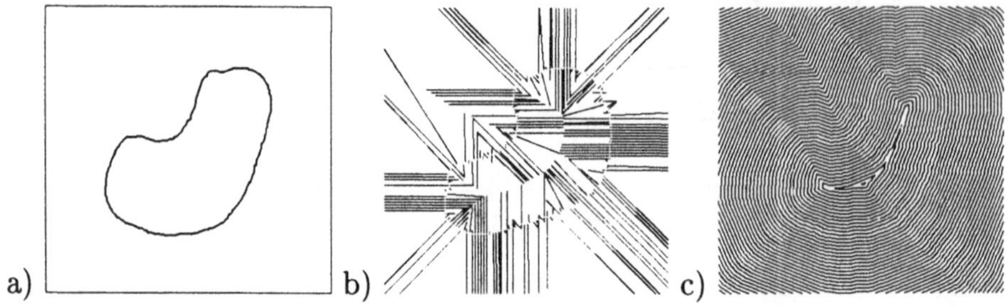

Figure 3.1. Level sets of the Voronoi Ordered Space: a) for a given contour C, b) $\phi_V(C, \vec{y}) = c$, level sets of the Voronoi Order, the first component of the Voronoi Ordered Space, c) $\delta_V(C, \vec{y}) = c$, level sets of Voronoi Distance, the second component of the Voronoi Ordered Space.

We define a mapping based upon a contour $C \in \mathbf{C}$ from the $\Re^2$ image space to another $\Re^2$ space.

DEFINITION 3.2 (VORONOI ORDERED SPACE) *Given a contour* $C \in \mathbf{C}$, *we define a mapping* $\Re^2 \rightarrow \Re^2$ *onto a* **Voronoi Ordered Space** *(see Figure 3.1), as follows:*

$$\vec{x} \Rightarrow \langle \phi_V(C, \vec{x}), \delta_V(C, \vec{x}) \rangle \quad (3.1)$$

$$\phi_V(C, \vec{x}) = \frac{1}{\|C\|} \min \left(\arg_s \min_{0 \leq s < \|C\|} (\|\vec{x} - \vec{\gamma}(s)\|) \right), \phi_V \in [0, 1) \quad (3.2)$$

$$\delta_V(C, \vec{x}) = \min_{\vec{y} \in C} (\|\vec{x} - \vec{y}\|), \delta_V \in [0, \infty) \quad (3.3)$$

i.e., for a given point, the mapping of Voronoi Ordered Space can be separated into two mappings $\Re^2 \rightarrow \Re$: the first (ϕ_V) is called the *Voronoi Order* and is determined by minimum value of the arc length parameter of the closest point on C; the second (δ_V) is called the *Voronoi Distance* and is the shortest distance from the point to the contour C. With this mapping, we define a new coordinate system where we replace the x-axis of our Cartesian coordinate system with a contour C. This pairing of Voronoi Order and Voronoi Distance creates a new description of the image space, warped w.r.t. the contour C. As shown in Figure 3.1, the level sets of the Voronoi Order, i.e., a partitioning of the space into sets of points of the same value, are the orthonormal projections of the contour C and the Voronoi Distances are the equidistant rings of C. If we overlay the level sets of Voronoi Order with level sets of the Voronoi

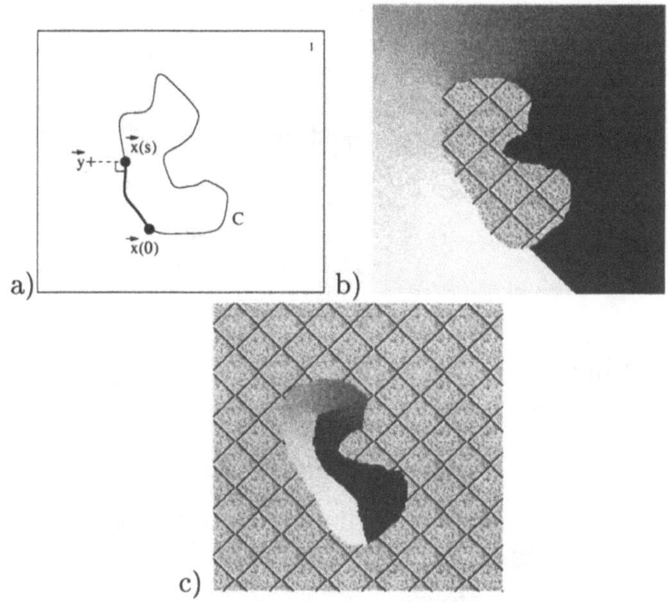

Figure 3.2. Voronoi Order of a contour C ∈ **C**: Given $\vec{y}$ on the image space, $\phi_V(C, \vec{y}) = \frac{s}{\|C\|}$ (see Definition 3.2) for a contour C b) the Voronoi Order on the exterior of C as image intensity can guide the object boundary search through order consistency in Chapter 4 (see Definition 3.7) c) the Voronoi Order on the interior of C; its discontinuities lead to a weighted skeleton called Voronoi Order Skeleton, used for MPEG-7 shape query in Chapter 6.

Distance, we create a warped grid where crossings between the two level sets are orthogonal.

By treating the image plane as a graph where each pixel is a node and the edges correspond to pixel adjacency, a good approximation of the Voronoi Ordered Space mapping can be calculated by a modified Dijkstra's shortest path algorithm. This algorithm runs in time $O(n \log n)$ where n is the number of pixels [Cormen et al., 1990].

4. PROPERTIES OF VORONOI ORDERED SPACE

This section proves some key properties of the Voronoi Ordered Space and gives some insights of how to apply Voronoi Ordered Space in our system designs. This section condenses proofs presented by Ogniewicz and Kubler [Ogniewicz and Kubler, 1995]. We focus upon the Voronoi Order, and how it orders the image space (see Figure 3.2). If we consider only contours that are monotonically increasing along its path w.r.t. the Voronoi Order of a given contour, we can find a subset of contours related

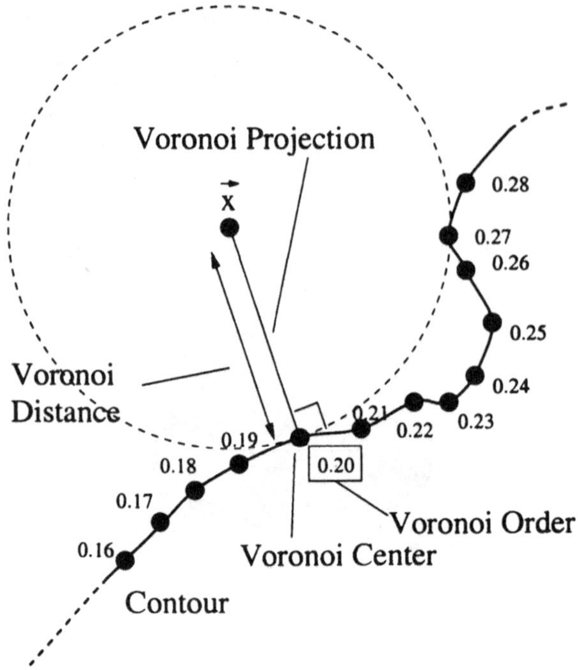

Figure 3.3. Terminology for our proofs: Voronoi Center, the minimally distant point on C of minimal arc length value to the given point $\vec{x}$; Voronoi Order, the arc length parameter of the Voronoi Center; Voronoi Distance, the distance from $\vec{x}$ to the Voronoi Center; Voronoi Projection, the line segment from $\vec{x}$ to the Voronoi Center

by shape to the given contour. Systems may integrate shape information by only considering this restricted set of contours.

To aid our analysis, we define two structures related to the Voronoi Order: *Voronoi Center* and *Voronoi Projection* (see Figure 3.3). Given a point $\vec{x}$ on the image plane, the *Voronoi Center* is the point on the contour C that is closest to $\vec{x}$ of minimum arc length value; the *Voronoi Projection* is the line segment from $\vec{x}$ to its Voronoi Center and its length is the Voronoi Distance.

THEOREM 3.1 (DISTINCTNESS OF MINIMAL DISTANCE PROJECTIONS) *Given a contour $C \in \mathbf{C}$ and two different points in $\Re^2$, $\vec{a}$ and $\vec{b}$, if the two points do not have the same minimally distant points on C, $\vec{y}$ and $\vec{z}$, respectively, then AY and BZ do not intersect.*

Proof by Contradiction. As shown by Figure 3.4, we assume the line segments AY,BZ intersect at a point $\vec{x}$. Without loss of generality, we assume that $|XY| \geq |XZ|$. By Schwartz Inequality,

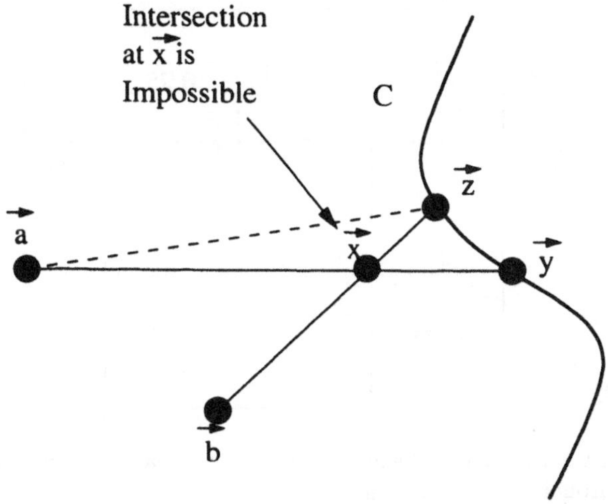

Figure 3.4. Theorem 3.1 (Distinctness of Minimal Distance Projections on C) Given two points $\vec{a}$,$\vec{b}$ and their respective minimal distance points on C, $\vec{y}$ and $\vec{z}$, we prove that the line segments AY and BZ cannot intersect.

$$|AY| = |AX| + |XY| \geq |AX| + |XZ| \geq |AZ| \tag{3.4}$$

Let us consider the two cases of $|AY| \geq |AZ|$. Case #1: if $|AZ| = |AY|$, then $\|AX\| + \|XZ\| = \|AZ\|$, $\|XZ\| = \|XY\|$ and $\angle AXZ = 180°$, forcing $\vec{y}$ and $\vec{z}$ to coincide and contradicting that $\vec{y}$ and $\vec{z}$ are distinct. Case #2: if $|AY| > |AZ|$, then $\vec{y}$ is not the minimally distant point on C to $\vec{a}$. ∎

COROLLARY 3.1 *(Distinctness of Voronoi Projections) Given a contour $C \in \mathbf{C}$ and two different points with different Voronoi Orders, the Voronoi Projections of the two points do not intersect.*

Proof. This corollary is a special case of Theorem 3.1. ∎

This result shows that the Voronoi Ordered Space acts as an alternative coordinate system: Voronoi Projections of different Voronoi Centers do not intersect and provide a consistent axis of the Voronoi Order for an alternative coordinate system.

THEOREM 3.2 (CONNECTIVITY OF VORONOI ORDER REGIONS) *Level sets of the Voronoi Order form connected regions.*

Proof. Consider any two points, $\vec{a}$ and $\vec{b}$, with the same Voronoi Order. Since they have the same Voronoi Order, they have the same Voronoi Center, $\vec{x}$. Consider the path of AX and XB. All points of AX and XB

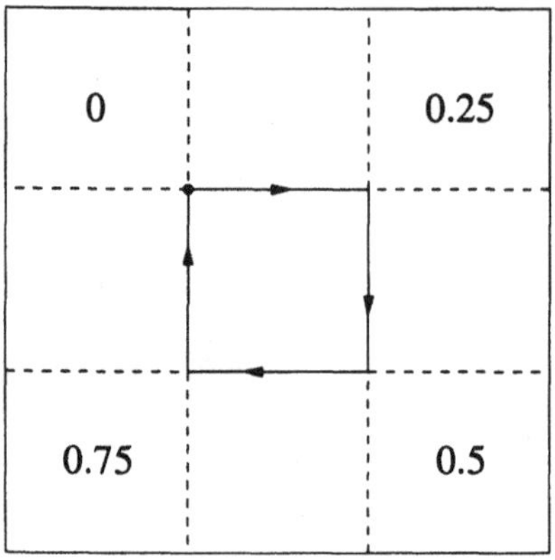

Figure 3.5. On the exterior of a square with corners, each corner induces a level set region of the Voronoi Order that is not a line.

have the same Voronoi Order. Therefore, any two points of the same Voronoi Order are connected by a path of the same Voronoi Order. ■

In Figure 3.1, the level sets of the Voronoi Order level sets are lines. However, as shown in Figure 3.5, if the contour is not smooth, the Voronoi Order may induce level sets that are connected regions, but not lines.

THEOREM 3.3 (VORONOI ORDER FOR SMOOTH CONTOURS) *Level sets of the Voronoi Order for a smooth contour are line segments of finite or infinite length.*

Proof. At a given point, a smooth contour has a unique tangent and hence up to orientation a unique normal $\vec{\nu}(s)$. Given two points, $\vec{x}$ and $\vec{y}$, that have the same Voronoi Center $\vec{z}$, and a contour C, since XZ and YZ are Voronoi Projections, they must both follow the same normal $\vec{\nu}(s)$ and therefore are collinear. ■

For a smooth contour, the level sets of a contour are straight line segments and the Voronoi Order can uniquely describe any point of the exterior (or interior) of the contour as a tuple of its Voronoi Order and Voronoi Distance.

In summary, these theorems establish the Voronoi Ordering Space as a *warped version* of 2-D image space. Given a contour $C \in \mathbf{C}$, the level sets of Voronoi Order partitions the image space into a series of connected regions. If we order these connected regions by their Voronoi Order value, we can consider a subset of contours in $\Re^2$ whose path moves from region to region while monotonically increasing Voronoi Order value. In the next section, we show that Voronoi Order has deep connections to shape structure.

5. DEFINITION OF THE VORONOI ORDERED SKELETON (VOS)

We derive a multi-resolution shape extraction technique for contours that is approximated well by calculations upon Voronoi Order.

DEFINITION 3.3 (VORONOI ORDERED SKELETON) *Given a contour $C \in \mathbf{C}$, and a point $\vec{y}$ on the interior of C, we define a function χ_V on $\Re^2$ called the Voronoi Order Skeleton (VOS):*

$$\chi_V(C, \vec{y}) = \begin{cases} \max_{\vec{a} \in S, \vec{b} \in S} \left(\rho(C, \vec{a}, \vec{b}) \right) & , \|S\| > 1 \\ 0 & , \|S\| = 1 \end{cases} \quad (3.5)$$

where S are the set of points on C that are minimally distant from $\vec{y}$, $\vec{a}$ and $\vec{b}$ are points in S, ρ is the minimal length of a segment of C that goes between $\vec{a}$ and $\vec{b}$, and $\|S\|$ is the number of points in S.

We can approximate the Eq. 3.5, using our previous definition the Voronoi Order, ϕ_V(Eq. 3.1):

$$\chi_V(C, \vec{y}) \approx \max_{\|\vec{u}\|=1} \left(\min \left(\begin{matrix} |\phi_V(C, \vec{y} + r\vec{u}) - \phi_V(C, \vec{y})|, \\ |1 - \phi_V(C, \vec{y} + r\vec{u}) + \phi_V(C, \vec{y})| \end{matrix} \right) \right), r << 1 \quad (3.6)$$

where r is the smallest resolution of the image and the minimization term in Eq. 3.6 is a mathematical expansion of ρ of Eq. 3.5 in terms of ϕ_V. Note that in the approximation, if $\hat{d}$ is the minimal distance from $\vec{y}$ to C, we find all the points on C that are within a distance $\left(\hat{\delta} \pm r\right)$ and we do not consider all pairs between minimally distant points on C, but between the Voronoi Center of $\vec{y}$ and all points on C that are within a distance $\left(\hat{d} \pm r\right)$. For high resolution images where r is small compared to the contour length, Eq. 3.6 is a good approximation, as shown in Figure 3.6. After calculating the Voronoi Order, this approximation of the VOS is computationally equivalent to an image convolution with a 3x3 matrix. Thus, the running time for calculating VOS is $O(n \cdot \log n + n)$ for the Voronoi Order calculation and maximization of Eq. 3.6 on every pixel over its adjacent pixels.

How do we use Voronoi Order Skeleton?

Voronoi Order Skeleton is a way of expressing shape.

Shape is inherent, descriptive property of a video object. For instance, we can classify the object on the left and infer its class, pose and activity by its outline.

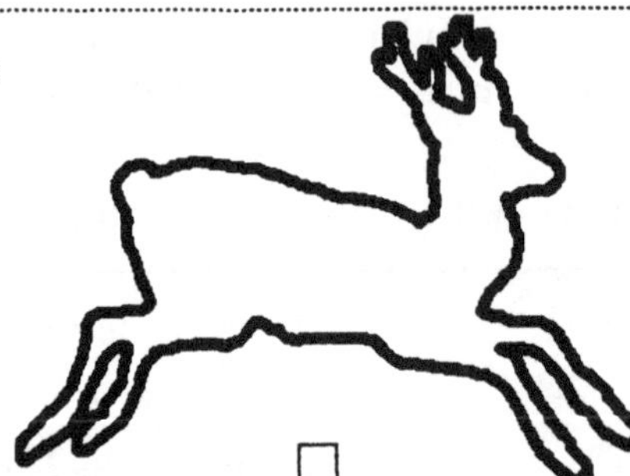

If we wish to use shape as a query, we produce a shape representation that selectively reduces information. For content query invariant of pose and activity, a skeleton-based representation fits well for selected subset of objects.

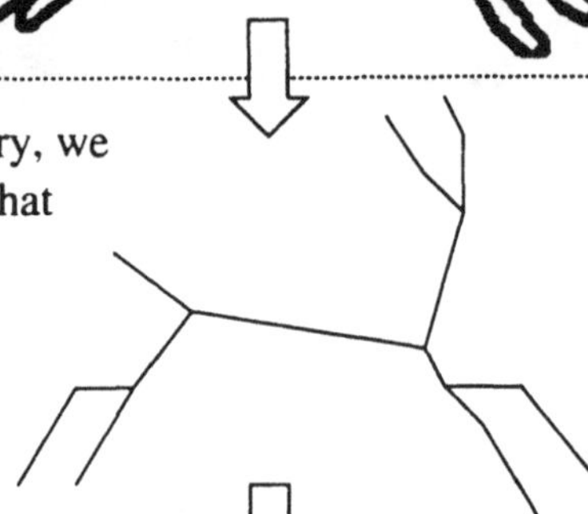

Comparing two skeletons is difficult because of the different ways of aligning two skeletons. However, the Voronoi Order Skeleton has an inherent hierarchy given by the structure of the shape.

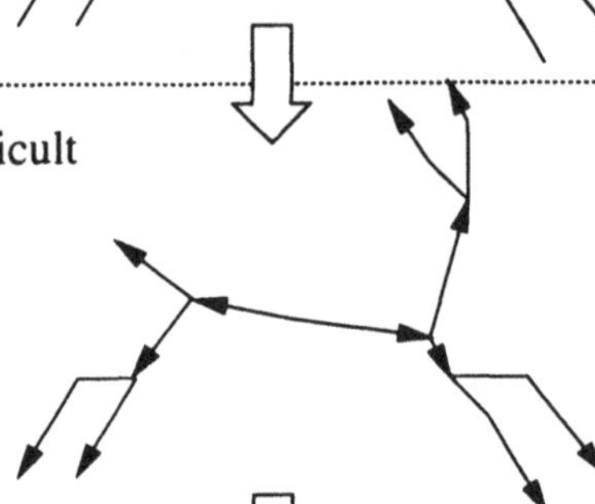

With Voronoi Order Skeleton, we abstract the shape in the form of an ordered tree. Comparisons of these trees will implement a content query invariant of pose and activity.

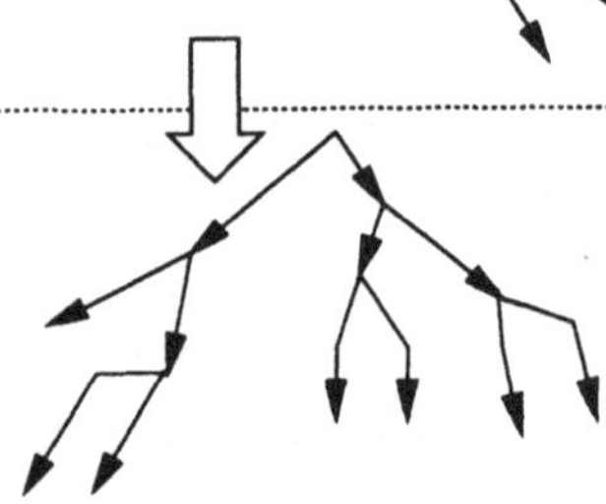

Voronoi Order Skeleton is a powerful tool for using object shape to recognize video content

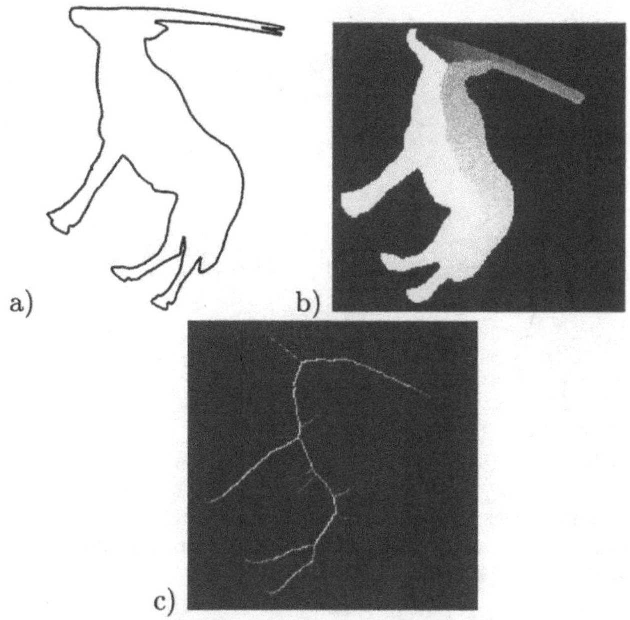

Figure 3.6. Steps in calculating the Voronoi Ordered Skeleton a) the Original Contour, b) the Voronoi Order of the interior where the image intensities are mapped to order values, c) the Voronoi Order Skeleton, calculated from the Voronoi Order (see Eq. 3.6).

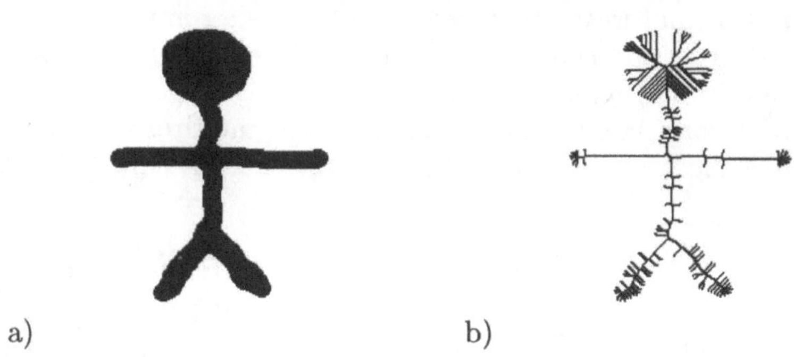

Figure 3.7. The Tree Structure of Voronoi Ordered Skeleton: given a contour in a), we can derive a Voronoi Order Skeleton in b). We can map the points on the interior to a tree where value of χ_V (Eq. 3.6) reflects at what depth the point exists within the tree.

6. MULTIRESOLUTION PROPERTY OF THE VOS

In this section, we show that χ_V is formally related to a multiresolution representation of shape contour. From analysis of the MAT, we

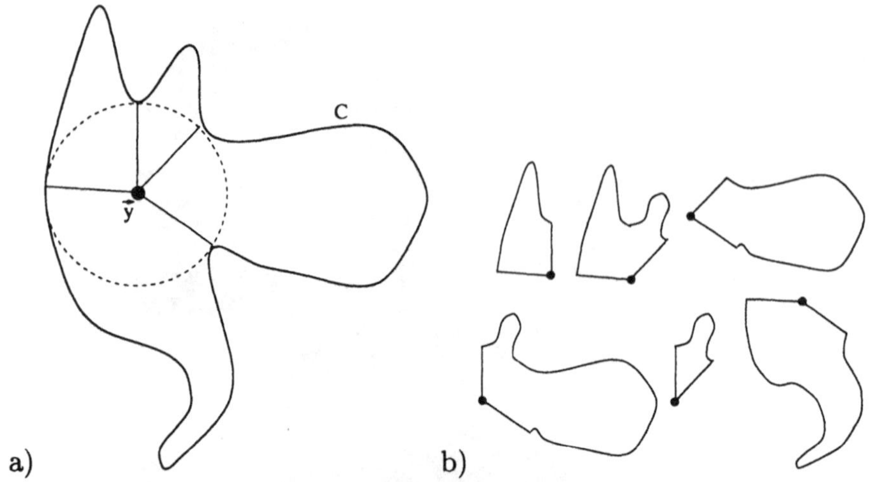

Figure 3.8. Voronoi Areas of a point $\vec{y}$,$\mathbf{\Phi}(\vec{y})$: If a) $\vec{y}$ is equidistant from four different distinct points on the contour C, then b) there are six different Voronoi Areas associated with $\vec{y}$.

know that this skeleton formed by the non-zero values of χ_V can be mapped to a graph that is a tree as shown in Figure 3.7. The analysis in this section shows the value of χ_V is related the depth of a point within multiresolution version of that tree: the χ_V value of a parent in the tree is always greater than or equal to value of its children. From χ_V values, we can derive this multiresolution tree and use it as a robust shape representation. This section condenses proofs presented by Ogniewicz and Kubler [Ogniewicz and Kubler, 1995].

To show the connection between χ_V and the multiresolution tree, we must define the underlying structures of the contour called Voronoi Areas:

DEFINITION 3.4 (VORONOI AREAS AND PIVOTS) *Given a contour $C \in \mathbf{C}$, a point $\vec{y}$ on the interior of C, let S be the set of points on C that are minimally distant from $\vec{y}$. As shown in Fig. 3.8, we define $\mathbf{\Phi}(C, \vec{y})$, the Voronoi Areas of $\vec{y}$, are the set of regions of finite area described by all pair-wise combinations of S through the following construction (for a pair of points $\vec{a}$, $\vec{b}$):*

1. *the line segment AY*
2. *the line segment BY*
3. *a segment of the contour C from $\vec{a}$ to $\vec{b}$ whose length is less than or equal to half the perimeter of the contour C.*

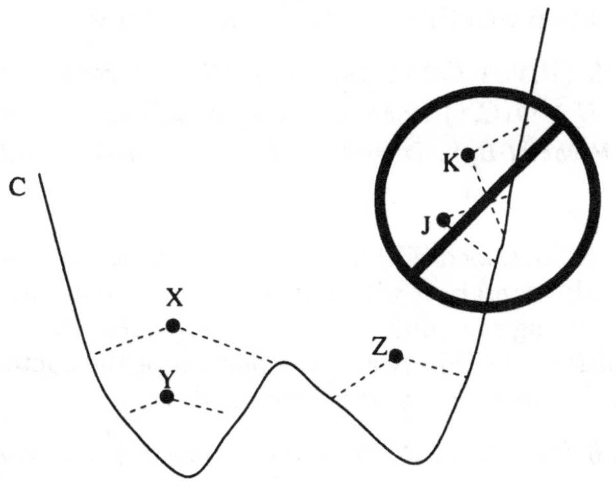

Figure 3.9. Relations between Voronoi Areas: As shown above, Voronoi Areas may completely contain one another as X does to Y, or be mutually exclusive as X is to Z, or as Y is to Z. However, if J and K are Voronoi Areas, J and K cannot intersect without one containing the other.

If $\|S\| > 1$, we call the point $\vec{y}$ the Voronoi Pivot for the Voronoi Areas in $\mathbf{\Phi}(C, \vec{y})$.

Next, we prove that two Voronoi Areas that have different pivots are related in one of the two ways: either they are non-intersecting or one contains the other.

THEOREM 3.4 (RELATIONS BETWEEN VORONOI AREAS) *Given a contour $C \in \mathbf{C}$, $A \in \mathbf{\Phi}(C, \vec{a})$, and $B \in \mathbf{\Phi}(C, \vec{b})$ where $\vec{a}$ and $\vec{b}$ are different Voronoi Pivots, A and B are only related one of two ways:*

1. *either A completely contains B or vice versa, i.e. ,*

$$(A \cap B = B) \,\mathbf{XOR}\, (A \cap B = A)$$

2. *or A and B are mutually exclusive w.r.t area, i.e. ,*

$$(A \cap B = \emptyset)$$

Proof by Contradiction. The proof uses the non-intersection properties of the theorem 3.1. To have an intersection of A and B that is neither A or B implies that a minimal distance projection of $\vec{a}$ intersects with one of minimal distance projections of $\vec{b}$. Since this is impossible, the statement must be true. ■

We now relate pivots through their Voronoi Areas.

THEOREM 3.5 (PIVOT CONTAINMENT) *Given a contour $C \in \mathbf{C}$, $A \in \mathbf{\Phi}(C, \vec{a})$, and $B \in \mathbf{\Phi}(C, \vec{b})$ where $\vec{a}$ and $\vec{b}$ are different Voronoi Pivots, if the Voronoi Pivot of B, $\vec{b}$, is contained in A, then B is contained within A.*

Proof. If $\vec{b}$ is contained within A, then the minimal distance projections of $\vec{b}$ are also contained within A, since they cannot go outside of A without intersecting the minimal distance projections of $\vec{a}$. Since all the boundaries of B are within A and the portion of the contour perimeter that B borders is less than $\frac{1}{2}$, B is contained by A.

THEOREM 3.6 (VORONOI PIVOT CONTAINMENT) *Given a contour $C \in \mathbf{C}$, $A \in \mathbf{\Phi}(C, \vec{a})$, and a Voronoi Pivot $\vec{b}$, if a Voronoi Pivot $\vec{b}$ is contained in the interior of Voronoi Area A and $A \notin \mathbf{\Phi}(C, \vec{b})$, every Voronoi Area in $\mathbf{\Phi}(C, \vec{b})$ is contained in A.*

Proof. Since every Voronoi Area in $\mathbf{\Phi}(C, \vec{b})$ has the same Voronoi Pivot $\vec{b}$ and $\vec{b}$ is contained in A, A contains all Voronoi Areas in $\mathbf{\Phi}(C, \vec{b})$. ∎

Finally, we need one definition in order to relate the values of χ_V with the Voronoi Areas.

DEFINITION 3.5 (VORONOI-CONTAINMENT FOR VORONOI PIVOTS) *A pivot $\vec{a}$ is said to Voronoi-contain a pivot $\vec{b}$ iff any Voronoi Area in $\mathbf{\Phi}(\vec{a})$ contains all Voronoi Areas in $\mathbf{\Phi}(\vec{b})$.*

With these definitions and theorems, we define the multiresolution property of this containment relation over interior of a contour $C \in \mathbf{C}$.

THEOREM 3.7 (TREE STRUCTURE OF VORONOI PIVOTS) *Given a contour $C \in \mathbf{C}$, the Voronoi-Containment relation over the Voronoi Pivots on the interior of C induces a forest, i.e., a set of trees. Each node in each tree is a pivot in the interior of C and the parent-descendant relation in each tree follows the Voronoi-containment of pivots.*

Proof. Let us consider all Voronoi Pivots on the interior of C as set I. These Voronoi Pivots can be divided trivially into two sets: points that are Voronoi-contained by another Voronoi Pivot and those that are not, S and $\bar{S}$, respectively. We partition pivots in I by which point in S they are Voronoi-contained. Let us consider one partition T, a set of these pivots. Voronoi Containment induces a tree on the set T, where a node is a member of T and the paths in the tree describe the parent-descendant relation that follows the Voronoi-containment relation because:

1. Every node is connected to the root.
   Every Voronoi Area is Voronoi-contained by one pivot in T.

2. There are no cycles in the graph.
   Two different Voronoi pivots that contain each other contradict theorem 3.6.

3. There are no reconverging paths.
   A reconverging path implies that there exists two different Voronoi Areas that do not contain each other, but contain a common pivot and, consequently, have a common area, contradicting theorem 3.4.

Therefore, Voronoi Containment on the set of points in T induces a tree. Applying this analysis to every partition, Voronoi Containment induces a forest over the Voronoi Pivots of C. ■

In Chapter 6, we wish to have a single rooted tree to represent a shape contour. For the case where forest contain multiple trees, we merely create another node and join all roots of trees to this node to create a tree with a single root. Although, in theory, theorem 3.7 implies that there may be multiple trees, in practice, after thousands of tree extractions, we have always found that the tree has only one root. We conjecture that the forest in theorem 3.7 is always a single tree. Although a proof of this conjecture would allow for a more elegant analysis, it is irrelevant for our applications and beyond the scope of this book.

The next step is to relate the containment structure of Voronoi Pivots and the χ_V values by defining a function that quantifies a measure of Voronoi-Containment.

DEFINITION 3.6 (THE SPAN OF A VORONOI AREA) *Given a Voronoi Area A, the spanning function $\alpha(C, A)$ returns the length of the perimeter of A that coincides with C.*

This function on Voronoi Areas allows the containment relation between Voronoi Areas to be expressed as a value relation.

THEOREM 3.8 (CONTAINMENT INDICATOR) *For a given contour C, two Voronoi Areas, A and B,*

$$A \text{ contains } B \rightarrow \alpha(C, A) \leq \alpha(C, B). \tag{3.7}$$

Proof. If Voronoi Area A contains B, then the perimeter of B that borders the contour C is also contained by the perimeter of A that borders C. ■

Now we may proceed to our final result.

THEOREM 3.9 (MULTI-RESOLUTION OF χ_V) *Given a contour $C \in \mathbf{C}$, the VOS values on the Voronoi Pivots are mapped onto a tree where the ancestor-descendant relationship is monotonically decreasing w.r.t. χ_V.*

Proof. Consider the Voronoi Ordered Skeleton value of the two pivots $\vec{a}$ and $\vec{b}$ in child-parent relationship where $\vec{a}$ is the parent and $\vec{b}$ is the child. Since $\vec{a}$ is a child, all members of $\mathbf{\Phi}(\vec{b})$, must be contained by a member of $\mathbf{\Phi}(\vec{a})$, $\hat{A}$. Using theorem 3.8,

$$\vec{a} \text{ Voronoi} - \text{Contains } \vec{b} \rightarrow \max_{B \in \mathbf{\Phi}\left(C, \vec{b}\right)} (\alpha(C, B)) \leq \alpha\left(C, \hat{A}\right) \tag{3.8}$$

Loosening the upper bound,

$$\max_{B \in \mathbf{\Phi}(C, \vec{b})} (\alpha(C, B)) \leq \alpha\left(\hat{A}\right) \leq \max_{A \in \mathbf{\Phi}(C, \vec{a})} (\alpha(C, A)) \tag{3.9}$$

This maximum operation is equivalent to our definition of χ_V.

$$\chi_V(C, \vec{a}) \leq \chi_V\left(C, \vec{b}\right) \tag{3.10}$$

By taking the contrapositive of Eq. 3.8, we can relate χ_V the Voronoi-containment between pivots.

$$\left(\chi_V(C, \vec{a}) > \chi_V\left(C, \vec{b}\right)\right) \Rightarrow \vec{b} \text{ does not Voronoi-contain } \vec{a} \tag{3.11}$$

i.e., if the VOS value of $\vec{a}$ is greater than the VOS value of $\vec{b}$, then the pivot $\vec{a}$ cannot be a parent of $\vec{b}$ pivot in the tree induced by the VOS. ∎

Coupled with the fact that the non-zero points of the VOS (the MAT), form a tree structure, χ_V can be used to derive a tree structure that reflects the Voronoi-containment relation of a given shape.

In summary, these theorems show that the Voronoi Order Space has a deep connection to shape structure. For all practical purposes, we can derive a singly rooted tree structure from the non-zero χ_V values of a given shape contour. Furthermore, the depth and branching of this tree structure are directly related to an implicit containment relation that can be directly derived from the shape of contour. Much like each parent in the tree that can represent all its descendants, the Voronoi areas from Voronoi Pivots contain all the areas derived from pivots within those areas. Each point in the tree structure can become a representation of all its children; likewise, the Voronoi Pivots related by Voronoi containment can become a multiresolution representation of shape. Using the VOS values, χ_V, we can derive this multiresolution representation and use it as a robust shape representation.

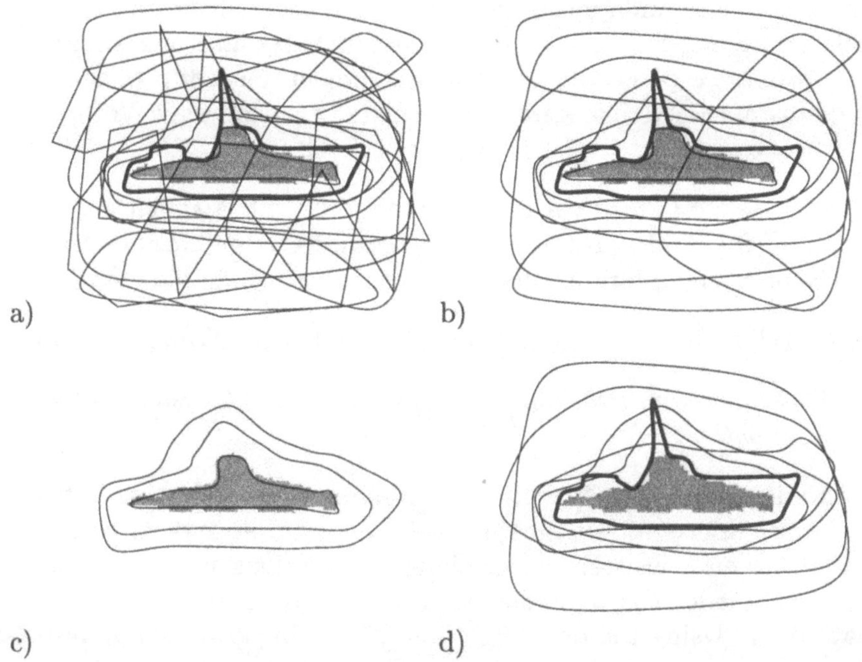

Figure 3.10. Levels of Contour Description: The initial estimate of shape is shown as the grey blob in the center of images; the true shape is shown as darkened contour. a) A set of curves, b) the subset of curves that are smooth, c) the subset of curves that are also equidistant, d) the subset of curves that are smooth, contains the initial estimate, and are Voronoi Order Consistent. Note that most restrictive set of curves that still contains the true contour is d).

7. APPLICATIONS

We use Voronoi Ordered Spaces in two applications in this book: Video Object Segmentation in support of the MPEG-4 standard and Image/Video Object Query by Shape in support of MPEG-7 standard.

APPLICATION #1: RESTRICTING OBJECT BOUNDARY SEARCH SPACE

In Chapter 4, the Voronoi Ordering provides a seamless interface of *a priori* shape information for our video object extraction system. As shown in Figure 3.10, if we have an reliable initial contour estimate, Voronoi Ordered Space can be used to restrict the search space of an object boundary search. In the most general and least restrictive case, we use smoothness. If we enforce that a boundary estimate must completely contain the initial estimate, we can further restrict our search space. If we assume our estimate has the correct shape of the object, we restrict the boundary estimate to be equidistant from the initial estimate, searching for only its size. Our reliability and precision of our initial estimate is precise enough for containment, but not precise enough for equidistance. We integrate this shape information via a Voronoi Ordered Space.

DEFINITION 3.7 (VORONOI ORDER CONSISTENT) *A simple closed contour* $C_\circ = \{\vec{y}_\circ(s) : 0 \leq s < \|C_\circ\|\}$, *is* **Voronoi Order Consistent** *with* C *iff there exists an arc length parameterization of* $C_\circ$ *such that*

$$\forall\{s_1, s_2 | 0 \leq s_1 < s_2 < \|C_\circ\|\}, \phi_V\left(C, \vec{y}_\circ(s_1)\right) < \phi_V\left(C, \vec{y}_\circ(s_2)\right) \quad (3.12)$$

i.e., $C_\circ$ *is consistent with Voronoi Order if* ϕ_V *along* $C_\circ$ *is monotonically increasing with* s.

For an initial contour (C) with an approximate shape, we can integrate this partial information through the Voronoi Order and its concept of Voronoi Order Consistency. Applying calculations upon this warped space, we restrict the boundary search space to guide a boundary estimate ($C_\circ$). Using the ordering criterion, we integrate shape into our video object segmentation problem. Given an initial estimate of our object boundary C, we aid our video object extraction in Chapter 4 by considering only a subspace of smooth contours that both enclose C and are Voronoi Order Consistent with C.

APPLICATION #2: SHAPE REPRESENTATION

In chapter 6, our solution for an object shape descriptor uses the Voronoi Ordered Skeleton as a mathematical description of a set of object assumptions called *implicit structure*, as defined in Section 6.3. Considering only shapes that are described by a simple closed contour, we can use the VOS as a robust representation of shape. We compare shapes through the mapping of their tree structures. After we address the problems of representing the continuous skeleton structure as a graph in Chapter 5, we can use VOS as a shape representation that can be invariant to particular bendings of a shape.

8. CONCLUSION

In this chapter, we have introduced the concept of Voronoi Ordered Space, a distillation of previous work that serves as an interface between shape information and our content-based processing systems. In Chapter 4, the concept of the Voronoi Ordered Space is used to integrate shape information into our video object extraction system. In Chapter 6, the concept of the Voronoi Order Skeleton derives a robust shape representation that is the basis of our shape query system. In Chapter 7, Voronoi Ordered Spaces can synergistically unify these two systems: representation supporting extraction and vice versa.

Chapter 4

A SYSTEM FOR VIDEO OBJECT SEGMENTATION

In this chapter, we implement a video object segmentation system that integrates Voronoi Ordered Spaces of Chapter 3 with existing techniques of Chapter 2 to support the MPEG-4 functionality of object-addressable video content [Lin and Kung, 1998a] [Lin et al., 1998] [Lin and Kung, 1999]. Our surface optimization formulation describes the video object segmentation problem (see Section 6.) in form of an energy function that integrates many visual processing techniques. By optimizing this surface, we balance visual information against predictions of models with *a priori* information and extract video objects from a video sequence. Since the global optimization of such an energy function is still an open problem, we decompose our formulation to allow a tractable optimization via dynamic programming in an iterative framework. In conclusion, we show the results of the system on the MPEG-4 test sequences and objectively compare them to results that are hand-segmented by the MPEG-4 committee.

1. PREVIOUS WORK

The MPEG-4 standard is the first major step to support content-based functionality (such as cut and paste, synthesis, and query of video objects) with its object-addressable content (see Figure 4.1). The major enabling technology for the MPEG-4 standard are systems that compute *video object segmentation*, i.e., the extraction of video objects from a given video sequence.

Research in video object segmentation has precursors in object tracking and detection technologies [Irani and Peleg, 1992] [Wang et al., 1995]. However, instead of extracting the video objects, object tracking only derives points on an object that represent the motion of the object as

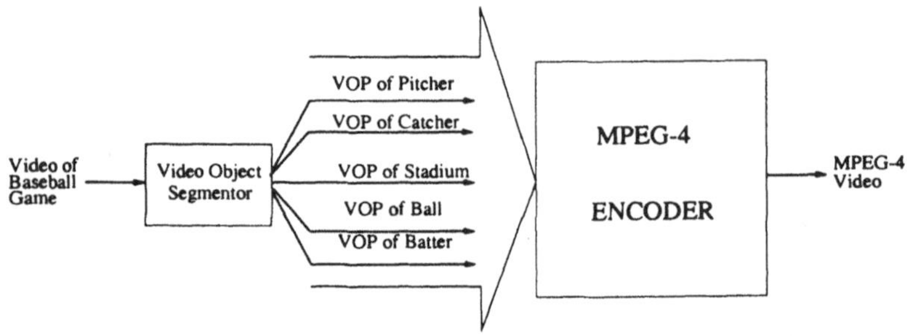

Figure 4.1. Role of Video Object Segmentation in MPEG-4: Video Object Segmentation is considered preprocessing before MPEG-4 encoding. The MPEG-4 encoder considers each Video Object Plane separately, but packages them into one video stream.

a whole and tracks them. Object detection finds only the number of objects within a scene. Object tracking and detection can contribute supporting preprocessing for our segmentation system.

There are extension of the tracking algorithms that use snakes to track the VOP boundary through time [Kass et al., 1988] [Gu and Lee, 1997] [Leymarie and Levine, 1993]. However, the energy minimization of snakes require an initial contour, usually provided by the user. In our own work, beyond the initial assumptions stated in Section 2., our process requires no human input and, as a result, is much more computationally intensive.

As noted in section 2.6, generalized Change Detection techniques analyze $\frac{\partial I}{\partial t}$ to determine the classification of pixels as foreground and background [Neri et al., 1998] [Til Aach, 1993]. These processes cannot differentiate between multiple objects and require further processing for video object segmentation. However, these approaches provide very powerful analyses that are also integrated into our system.

Many video object segmentation techniques use only motion information to extract video objects [Naseri and Stuller, 1996] [Adiv, 1985] [Bouthemy and Francois, 1993] [Meier and Ngan, 1998]. The simple clustering of motion vectors often extract accurate, but imprecise video objects. Near the video object boundary, motion analysis is often unreliable and not precise enough to derive high resolution video objects. In section 4., we use our own analysis from motion estimates as a bootstrap step for our system. For high resolution results, our system also integrate visual information (such as edges) as well as motion into its analysis.

Other relevant techniques are watershed analysis and mesh-based analysis. Watershed is related to our surface optimization algorithm through

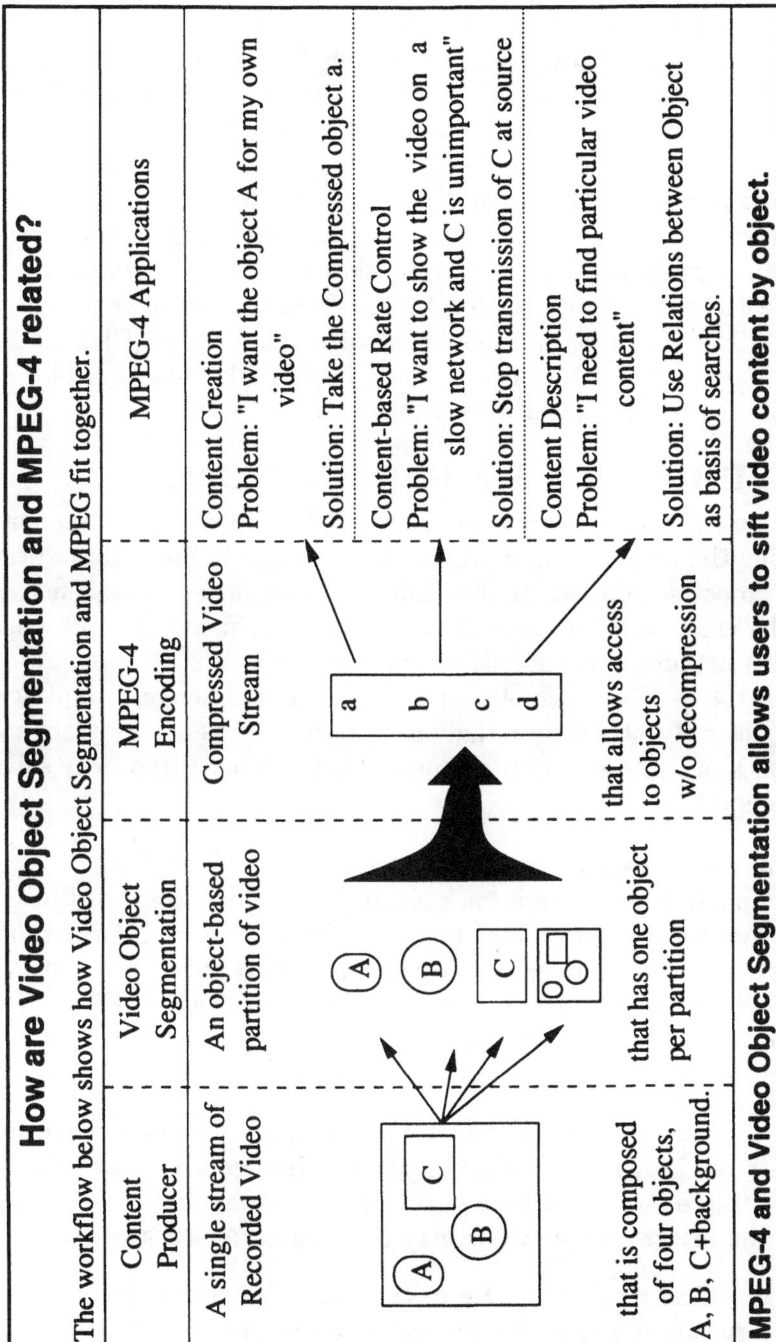

How are Video Object Segmentation and MPEG-4 related?
The workflow below shows how Video Object Segmentation and MPEG fit together.
Content Producer
Video Object Segmentation
MPEG-4 Encoding
MPEG-4 Applications
A single stream of Recorded Video
An object-based partition of video
Compressed Video Stream
Content Creation
Problem: "I want the object A for my own video"
Solution: Take the Compressed object a.
Content-based Rate Control
Problem: "I want to show the video on a slow network and C is unimportant"
Solution: Stop transmission of C at source
Content Description
Problem: "I need to find particular video content"
Solution: Use Relations between Object as basis of searches.
A
B
C
a
b
c
d
that is composed of four objects, A, B, C+background.
that has one object per partition
that allows access to objects w/o decompression
MPEG-4 and Video Object Segmentation allows users to sift video content by object.

dual problem representation [Wang, 1995]. However, it is unclear how this algorithm extends multiple frames. Mesh-based analysis divides the frame into a mesh of small triangular regions and does individual tracking of each region [Altunbasak et al., 1997]. Although the work tracks the object membership of each region well, the initial determination of membership of each region is not a well-understood problem.

In the current literature, Choi's work in video object segmentation is the most impressive in its theory and system design. His system design begins with the usual motion-based segmentation algorithm, progresses to the integration of visual information and, finally, expands the modeling of the motion information by a hierarchy of motion models [Choi et al., 1995] [Choi et al., 1997b] [Choi et al., 1997a] [Choi and Kim, 1996]. Our own work has progressed in a similar manner, but with a different type of computational framework.

2. PROBLEM SIMPLIFICATIONS

We simplify some aspects of the MPEG-4 video object segmentation to define the scope of our analysis. The general problem description for video object segmentation is as follows: given a video sequence, partition the video sequence into video objects, i.e., such that each partition contains the projection of only one physical object (see Figure 4.2) [Committee, 1998] [Wang and Adelson, 1994]. The following simplifications clearly classify the future challenges either 1) as an inherent system drawback or 2) as a system extension that can be temporarily solved by human guidance:

1. **The video sequence is grayscale.**
   Although color information may help to distinguish the objects from background, we only treat grayscale images. Color information is an extension of the dimensionality of this feature space. Exploiting color for segmentation involves the study of perceptual qualities of colors and is beyond the scope of this book. In the future, color or texture analysis may integrated into our system in Eq. 4.5.

2. **The video sequence contains the same objects throughout its run, i.e., no object completely leaves or enters the video sequence.** For our system, we rely upon manual object detection. In the future, this analysis may be done automatically.

3. **The video objects to be extracted have boundaries that are not primarily defined by the camera frame.** The background objects (such as the sea and land in Figure 4.2b) have different characteristics than foreground objects and should be treated as sepa-

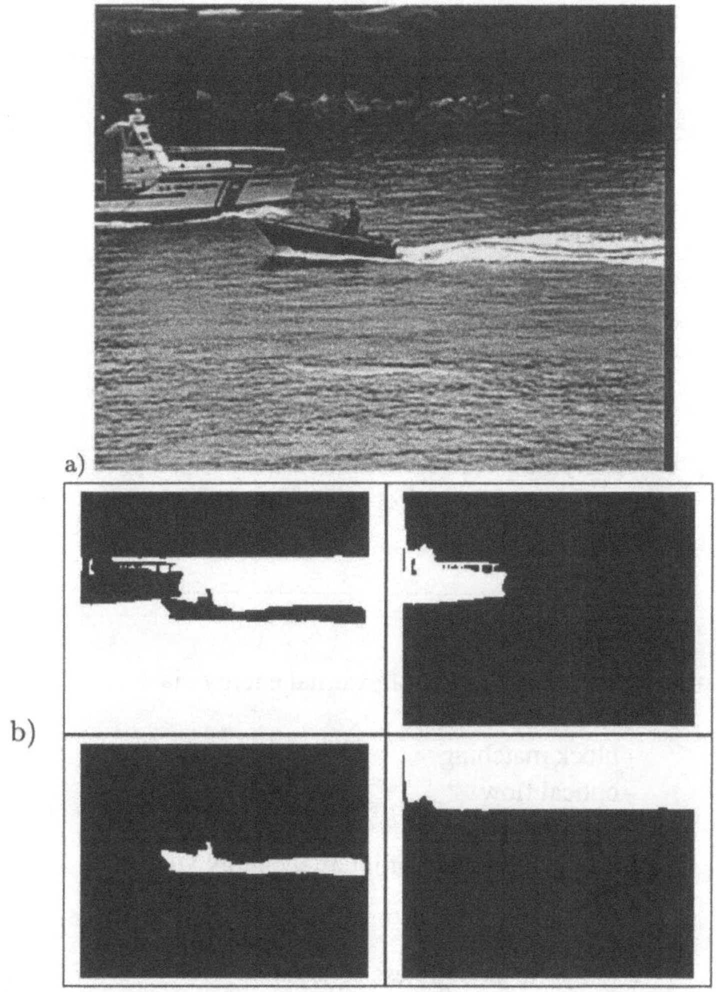

Figure 4.2. Ideal Video Object Segmentation Results: from the original video sequence as shown in a), the ideal video object segmentation system would divide the scene into b) four objects (in white, starting at the top right and going clockwise: ocean, large coastguard boat, small motorboat with wake and the shore). These results were the official hand-segmented results from MPEG committee of frame 30 from the coastguard sequence.

rate problems. We envision our system working with a background extraction system: the background extraction systems can then be applied to extract the imprecise background regions and our high resolution results of foreground objects would be subtracted from the background video objects to obtain precise background video objects. Thus, in this chapter, we primarily consider foreground objects.

How do We Extract Video Objects?

Our method is divided into Three (3) Major Steps:

1. FORMULATE

- Mathematically describe the Video Object as a surface
- Quantify how well the surface fits to visual artifacts of the video sequence as an "external energy" function
- Quantify how well the surface agrees with a priori knowledge of the object as an "internal" energy function

2. EXTRACT

- Map the quantification of external energy via extraction techniques:
  - block matching
  - optical flow
  - edge detection
- Map the quantification of internal energy via heuristics
  - spatial locality
  - temporal smoothness constraints
  - the Voronoi Order of the object shape estimate

3. OPTIMIZE

- Balance the two energy functions via Surface Optimization algorithm
  - A Dynamic Programming Algorithm within an iterative framework.

In this way, our Video Object Extraction algorithm manage many sources of information within a unified framework.

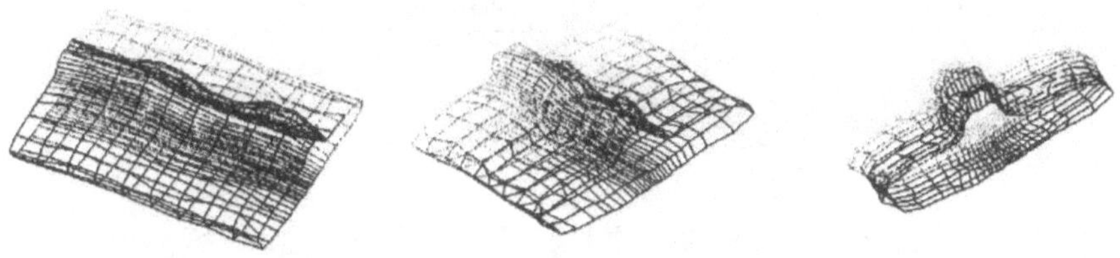

Figure 4.3. Surface-Representation of our video object: three different perspectives on the surface-based representation of video object of the large coastguard ship from the coastguard sequence frames 90-99.

4. **The objects do not have a high degree of occlusion.** Since we do not model interactions between objects, our analysis cannot handle a high degree of occlusion, i.e., when one object covers another from view. In the future, inter-object relationships will be an important aspect of the visual processing problems for both description and extraction.

3. PROBLEM FORMULATION VIA SURFACE OPTIMIZATION

The surface optimization framework provides a convenient mathematical formulation that can integrate many sources of information and is not bound to a particular computing architecture. We first treat the case of a single video object in isolation. We decompose the multiple video object segmentation into a set of isolated video object segmentations of single video objects. We identify and quantify our sources of segmentation information and formulate them in terms of surface optimization.

SURFACE VS. VOLUME REPRESENTATION

As shown in Figure 4.3, we represent the video object as a 2-D surface in the video space. Like the image segmentation problem in Chapter 2, there are two ways to represent a video object from a graph-theoretic perspective: either 1) volume-based as a partition of the 3-D dimensional spaces, or 2) surface-based as a set of interlocking surfaces. Although both exist in 3-D video space, a volume-based approach is effectively a 3-D problem while a surface is 2-D. With optimization techniques, the quality of the solution is often related to the number of parameters to optimize. A surface-based description is generally simpler because its behavior requires fewer parameters. Furthermore, by decomposing

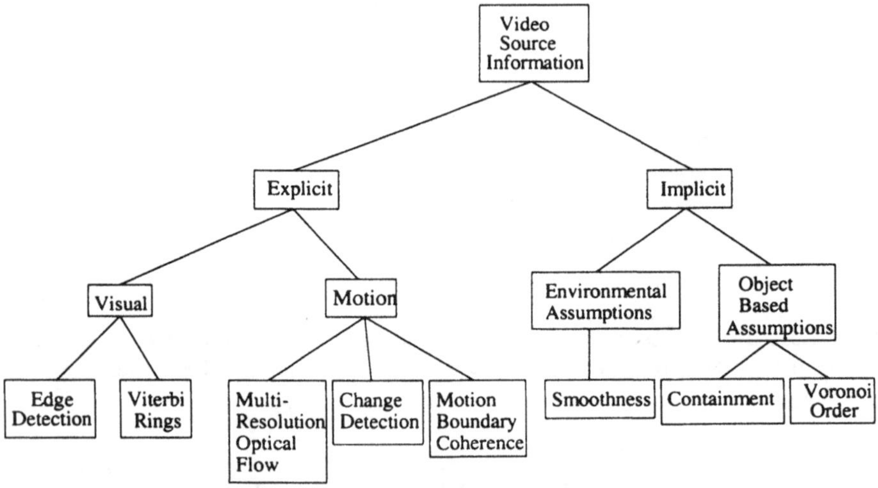

Figure 4.4. Taxonomy of visual information,subdivided by class and by energy component.

a surface down to a series of 1-D (contour) optimizations, we can apply dynamic programming. A volume-based approach does not have a similar decomposition. We choose a surface-based representation over a volume-based one because of the reduction of dimensionality.

NOTATION

Since this chapter will be referring to a surface and various decompositions of the surface, we define a series of notations to aid analysis:

1. $\mathcal{S}$ refers to a surface that represents a video object.
2. when the surface is within an iterative computation, $\mathcal{S}^{(n,t)}$ refers to the tth timeslice of the surface at iteration n.
3. when the surface is within an iterative computation, $\mathcal{S}^{(n,*)}$ refers to the surface at iteration n, i.e., $\mathcal{S}^{(n,*)} = \bigcup_t \mathcal{S}^{(n,t)}$.
4. when the surface is **NOT** within an iterative computation, $\mathcal{S}^{(*,t)}$ refers to the tth timeslice of the surface.
5. Vectors $(\vec{a}, \vec{b}, \ldots)$ are in $\Re^3$ on the video space.

ENERGY FUNCTION TAXONOMY

We organize our sources of segmentation information into a energy function taxonomy (see Figure 4.4). We define an energy function as a fitness metric for the surface representation of a video object and formulate the video object segmentation problem as follows:

$$\arg_{\mathcal{S}} \min \left(E_{vobject}(\mathcal{S}, I, M) \right) \tag{4.1}$$

where $E_{vobject}$ is an energy function, $\mathcal{S}$ is the spatio-temporal surface that represents the video object, I is the video sequence, and M is the *a priori* knowledge of the object. This energy function has a lower energy when the surface is consistent with I and M. At the highest level, we split the information into two basic classes: *internal* and *external* energies, $E_{internal}$ and $E_{external}$, respectively.

$$E_{vobject}(\mathcal{S}, I, M) = E_{internal}(\mathcal{S}, M) + E_{external}(\mathcal{S}, I) \tag{4.2}$$

External energies are deductive, i.e., the video object membership is deduced from visual artifacts. Internal energies are inductive, i.e., an assumption about the video object is made and the surface tries to fit the assumption. As shown in Eq. 4.2, these components differ in their functional arguments: internal energies are only dependent upon the surface and model descriptions; external energies are dependent on the surface and the video sequence. Internal energies counterbalances the external energy's tendency to fit (and overfit) the visual artifacts in the video sequence.

External and internal energies can be further classified by their time dependence and applicability, respectively. External energies can be split into two components, E_{image} and E_{motion}, based upon whether they use temporal correlation in their analysis. E_{image} does its analysis piecemeal, i.e., as if each frame is in isolation, while E_{motion} treats the video sequence as a whole in its analysis.

$$E_{external}(\mathcal{S}, I) = \int E_{image}(\mathcal{S}^{(*,t')}, I\left(x, y, t\right)|_{t=t'})dt' + E_{motion}(\mathcal{S}, I) \tag{4.3}$$

where $I\left(x, y, t\right)|_{t=t'}$ is the frame t' of the video sequence. Internal energies are split by their dependence upon *a priori* information: E_{world}, that are based upon environmental assumptions and are valid for all video objects, and E_{object}, that are based upon an assumption of object class:

$$E_{internal}(\mathcal{S}, M) = E_{world}(\mathcal{S}) + E_{object}(\mathcal{S}, M) \tag{4.4}$$

While environmental assumptions do not require any information about the object, their generality usually limits their power. Object-based assumptions are more powerful, but require the preprocessing to supply the applicable model and/or partial object information.

E_{IMAGE}

E_{image} expresses the surface attraction toward visual cues that mark the video object boundary in a given isolated frame. A general form of E_{image} is shown below:

$$E_{image}(\mathcal{S}^{(*,t')}) = \int_{\mathcal{S}^{(*,t')}} -R\left(I\left(x,y,t\right)|_{t=t'}, \vec{s}\right) ds \qquad (4.5)$$

where the function R returns the likelihood that an video object boundary exists at a given position $\vec{s}$ from the intensities of the frame t'. In our system, E_{image} is composed of two analyses: edge detection, the core technology that finds spatial image discontinuities, and Viterbi Rings, a refinement for edge detection that uses edge connectivity.

1. **Edge Detection**
   The primary cue for object boundary detection are edges. A surface that lies upon many edges is likely to be a good object boundary estimate. In our system, a Canny edge filter with a built-in edge follower suffices for grayscale images [Canny, 1986], although future work can generalize the concept of an edge to include color and texture.

2. **Viterbi Rings**
   Since all edges derived from edge detection do not necessarily belong to an object boundary, further analysis can distinguish between edges that are on the object boundary and those that are not. Given a known object center estimate, edges on object boundary tend to belong to a closed contour that go around the object center and to be oriented such that they face to the object center. We call this subset of edges with these properties, *Viterbi Rings.* However, we shall defer the determination of Viterbi Rings to the implementation (see Section 4.13) since its analysis also introduces important concepts for our surface optimization.

E_{MOTION}

E_{motion} expresses the attraction of the surface to discontinuities in the motion field. The surface S has a lower energy when it lies upon many motion discontinuities. E_{motion} has a general form in Eq. 4.6 below:

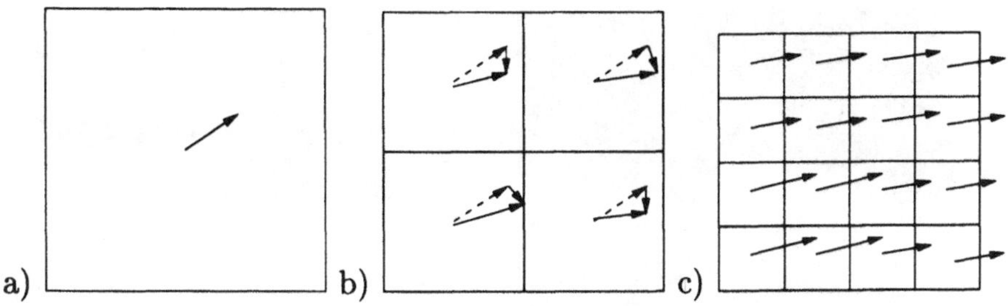

Figure 4.5. Iterations of multiresolution optical flow for progressive motion estimation a) At a low resolution, we do a optical flow algorithm over resolution block b) we split the low resolution motion information block and use it as initial conditions for the next resolution, c) we find the motion vectors at the higher resolution. We repeat the process and increase resolution until we reach pixel-level resolutions.

$$E_{motion}(\mathcal{S}) = \int_{\mathcal{S}} -g_{motion} \left(\begin{array}{c} \left(\frac{\partial v_x}{\partial x}\right)^2 + \left(\frac{\partial v_y}{\partial x}\right)^2 + \\ \left(\frac{\partial v_x}{\partial y}\right)^2 + \left(\frac{\partial v_y}{\partial y}\right)^2 \end{array} \right) dA \qquad (4.6)$$

where the projected motion field is $\vec{v}(x, y, t) = \langle v_x(x, y, t), v_y(x, y, t) \rangle$ and g_{motion} is a thresholding function. In our system, E_{motion} consists of three extraction techniques:

1. **Multiresolution Optical Flow**
   We use a multiresolution version of the original Horn and Schunck algorithm. Displacements of more than 2 or 3 pixels can cause the updating scheme of the original Horn and Schunck algorithm to fail. A multi-resolution pyramidal scheme can progressively estimate the motions at increasing resolutions (see Fig. 4.5) and use the motion estimates of the previous resolution to guide calculations of the current resolution. The motion field may be calculated in three successive passes at resolutions of $\frac{1}{4}, \frac{1}{2}$ and 1, expanded to the next resolution, and used as a initial conditions for the motion field of higher resolution. This process allows the optical flow to expand its range of maximum displacement from approximately 3 to 12 pixels.

   Motion clustering of this optical flow motion field can localize video objects and approximate the video object shape. As mentioned in Section 2.4, this optical flow algorithm is inaccurate near the object boundaries, although it gives good qualitative estimates of an object with respect to its motion. We use the optical flow to initialize our optimization and localize our objects. However, for precise determination of the object boundary, we combine our image analysis

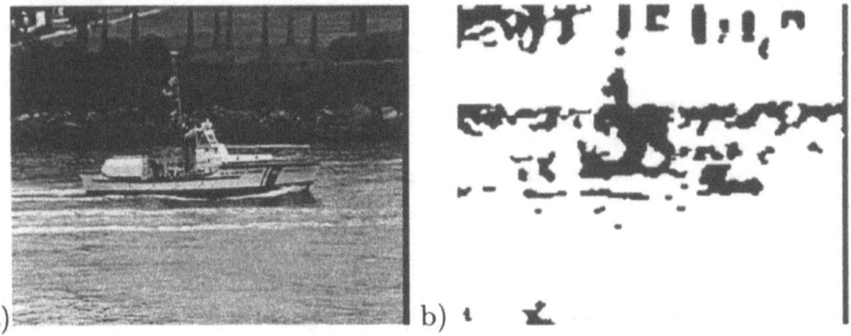

Figure 4.6. Change Detection Results: for a) frame 90 of coastguard sequence, we apply our change detection algorithm after majority motion compensation to extract the changing regions as shown in b). Note that homogeneous regions are not marked as changing, due to the aperture problem (of Chapter 2). Some background objects are marked as changing due to parallax.

techniques of section 4.4 with these motion estimates as our third component of E_{motion}.

2. **Change Detection**
   As mentioned in Section 2.6, change detection can identify general regions of change and aid in analysis (see Figure 4.6). Our change detection has a simple camera motion compensation that uses the majority motion vector in the frame. After compensating for camera motion, we use a simple threshold of variance on $\frac{\partial I}{\partial t}$ to find changing regions. Change detection requires a smaller smoothness neighborhood than motion estimation algorithms and localizes video object boundaries more precisely. Change detection also allows better discrimination between object and background edges.

3. **Boundary Motion Coherence**

   Using edge detection, we can enhance our motion estimation near object boundaries to find motion discontinuities. Assuming motion discontinuities coincide with edges and an edge is on an object boundary, an edge is **boundary motion coherent** (see Fig. 4.7) iff

$$(\vec{v_i} \neq \vec{v_o}) \wedge (\vec{v_i} \cdot \vec{N_e} = \vec{N_e} \cdot \vec{v_e}) \wedge (\vec{v_o} \cdot \vec{N_e} \neq \vec{N_e} \cdot \vec{v_e}) \qquad (4.7)$$

   where $\vec{v_i}$,$\vec{v_o}$,$\vec{v_e}$ and $\vec{N_e}$ are the motions of the inside, outside and edge regions, and the normal to edge, respectively. The interpretation of the boundary motion coherence test can be shown in Figure 4.8,

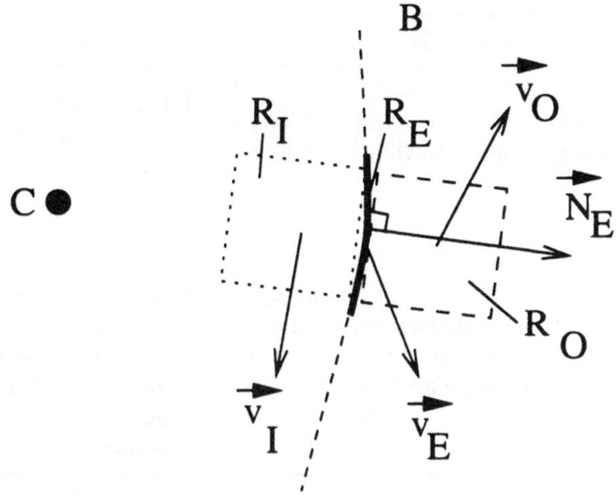

Figure 4.7. Boundary Motion Coherence: R_I, R_O, and R_E are the inside, outside and edge regions, respectively; $\vec{v_i}$,$\vec{v_o}$,$\vec{v_e}$, their respective motions; C, the object center; B boundary estimate; $\vec{N_e}$ normal to edge.

but, in short, we only enhance an edge if its interior moves with the edge and its exterior moves differently [Heitz and Bouthemy, 1993]. Although this test is very robust, it is computationally expensive and we cannot apply this test to all edges on the image. To solve this problem, we apply the boundary coherence tests to a limited set of edges, the Viterbi Rings. Note that Eq. 4.7 is independent of the motion estimation technique that is used. In our system, we use both block matching techniques and multi-resolution optical flow results in the boundary motion coherence tests. For small patches of highly textured regions, block matching techniques are more accurate than the optical flow, while for larger smoother regions, optical flow may resolve the projected motion that block matching cannot.

E_{WORLD}

E_{world} are internal energy components that requires no classification or *a priori* knowledge, i.e., they apply to all objects. Our system uses spatial and temporal smoothness as its only component of E_{world}. Other possible components in the same class of internal energies are Kalman filters that model Newton's Laws of Motion.

1. **Smoothness**
   For our smoothness constraint, we quantify both temporal and spatial smoothness of our surface by the magnitude of the second order gradient over the surface $\mathcal{S}$:

	$(\vec{v_i} = \vec{v_o})$	$(\vec{v_i} \neq \vec{v_o})$
$(\vec{v_i} \cdot \vec{N_e} \neq \vec{N_e} \cdot \vec{v_e})$ $(\vec{v_o} \cdot \vec{N_e} \neq \vec{N_e} \cdot \vec{v_e})$	Likely noise or possibly only the edge (not R_I) belongs to the object in question.	Unknown; edge, inside and outside region are not correlated.
$(\vec{v_i} \cdot \vec{N_e} = \vec{N_e} \cdot \vec{v_e})$ $(\vec{v_o} \cdot \vec{N_e} = \vec{N_e} \cdot \vec{v_e})$	Intra-object boundary; internal edge; edges have motion consistent with both inside and outside region.	Possible intra-object boundary or object boundary; uncertainty due to aperture problem w.r.t. the edge; edges could be motion-consistent with both inside and outside region.
$(\vec{v_i} \cdot \vec{N_e} \neq \vec{N_e} \cdot \vec{v_e})$ $(\vec{v_o} \cdot \vec{N_e} = \vec{N_e} \cdot \vec{v_e})$	Impossible.	Inter-object boundary from another object; edge moves with outside region, but not with inside region.
$(\vec{v_i} \cdot \vec{N_e} = \vec{N_e} \cdot \vec{v_e})$ $(\vec{v_o} \cdot \vec{N_e} \neq \vec{N_e} \cdot \vec{v_e})$	Impossible.	**Inter-object boundary**; edge moves with inside region, but not with outside region.

Figure 4.8. Truth Table for Boundary Motion Coherence. Note we only consider a an edge to be in the object boundary when $(\vec{v_i} \neq \vec{v_o}) \wedge (\vec{v_i} \cdot \vec{N_e} = \vec{N_e} \cdot \vec{v_e}) \wedge (\vec{v_o} \cdot \vec{N_e} \neq \vec{N_e} \cdot \vec{v_e})$ is true.

$$E_{world} = E_{smooth}(\mathcal{S}) = \int_{\mathcal{S}} \left\| \nabla^2 \mathcal{S} \right\|^2 dA \qquad (4.8)$$

where $\nabla^2 \mathcal{S}$ is the second order gradient of the surface $\mathcal{S}$.

E_{OBJECT}

Components of E_{object} use *a priori* knowledge and/or preprocessing results to predict surface shape and size. By knowing the approximate class of objects (for instance, rigid vs. non-rigid), we better predict the spatio-temporal behavior of the surface. In our system, our bootstrap stage gives approximate shape and size of an video object. As shown in Figure 4.9, we can translate this partial information through the concepts of containment, locality, and Voronoi Order consistency.

$$E_{object}(S, M) = E_{contain}(S, M) + E_{local}(S, M) + E_{voronoi}(S, M) \qquad (4.9)$$

Figure 4.9. Using Containment, Voronoi Ordered Space in the Surface Optimization: a) Original Frame, b) Bootstrap Estimate (textured region) with its Voronoi Order (gradient region), we search only contours that 1) contain the bootstrap estimate, 2) stay within some distance to the Bootstrap Estimate and 3) follow the gradient of Voronoi Order c) the optimal contour from this set

1. **Containment**
   We force our surface estimates to contain an initial conservative surface estimate.

$$E_{contain}(S, S_{initial}) = \int_S g_{contain}(\vec{x})d\vec{x} \tag{4.10}$$

$$\text{where } g_{contain}(\vec{x}) = \begin{cases} \infty, \vec{x} \text{ is inside } S_{initial} \\ 0, otherwise \end{cases},$$

   S is the surface to be estimated, and $S_{initial}$ is the initial conservative surface estimate.

2. **Locality**
   We force the surface estimate to be within a certain distance of an initial conservative surface estimate.

$$E_{local}(S, S_{initial}) = \int_S g_{local}(\vec{x})d\vec{x} \tag{4.11}$$

$$\text{where } g_{local}(\vec{x}) = \begin{cases} \infty, & \begin{array}{c} \vec{x} \text{ is further than distance} \\ \bar{d}(S_{initial}) \text{ from } S_{initial} \end{array} \\ & 0, otherwise \end{cases},$$

   S is the surface to be estimated, $S_{initial}$ is the initial conservative surface estimate, and $\bar{d}$ is a function that returns the distance dependent on $S_{initial}$.

3. **Voronoi Ordered Space**
   With a conservative estimate of our VOP, we restrict the search for object boundary estimate via Voronoi Order Consistency (see Section 7.). The Voronoi Order encodes shape information through

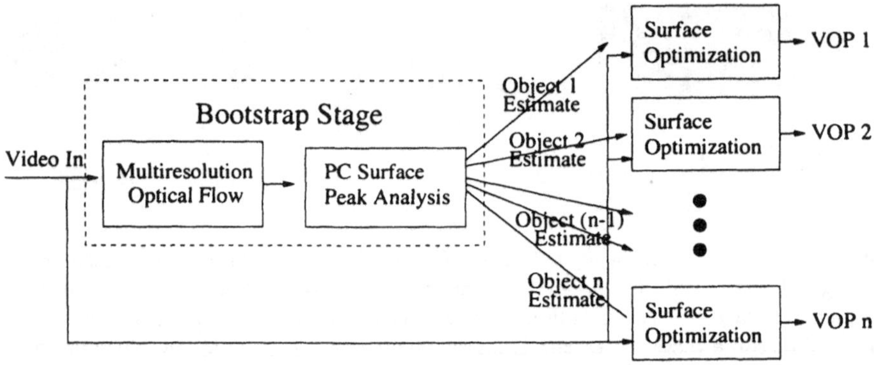

Figure 4.10. System Design: The bootstrap stage divides the multiple object segmentation into a series of isolated single object problems that can be solved through surface optimization.

warping the video space; the Voronoi Order Consistency ensures a contour estimate follows this order. Given an initial estimate of our object boundary C, we consider only surfaces whose timeslices are contours that are ordered consistently w.r.t. Voronoi Order.

$$E_{voronoi}(S, S_{initial}) = \int_t g_{voronoi}\left(S^{(*,t)}, S^{(*,t)}_{initial}\right) dt \tag{4.12}$$

$$\text{where } g_{voronoi}(A,B) = \begin{cases} 0, & \text{contour A is Voronoi Order consistent with contour B} \\ \infty, & otherwise \end{cases},$$

and the contour $S^{(*,t)}_{initial}$ is the timeslice at time t of the initial conservative surface estimate.

Note that the implementations of containment, locality, and Voronoi Ordered Space components are all hard constraints and a good surface estimate is highly dependent on that fact that $S_{initial}$ is conservative and correct. A probabilistic implementation of these E_{object} components would be more robust to inaccuracies in initial estimates.

4. THE BOOTSTRAP STAGE

We now address certain pragmatic issues associated with the surface optimization process:

1. **Initialization**
   As with most complex iterative optimizations, our system is highly dependent on initial values for good results.

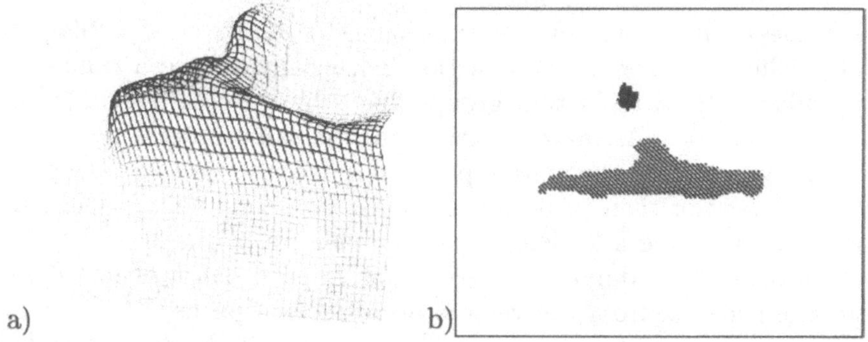

Figure 4.11. Principal Component Surface Analysis: a) by taking the first principal component of the motion paths of the video sequence with the initial point positions, we can derive a functional surface that represents the general movement of objects through the video sequences. b)We can derive approximate object positions and size through peak analysis of this surface: the bulk of coastguard ship and part of its mast. The results were taken from frames 90-99 of the coastguard sequence.

2. **Multiple Object Separation**
   Our formulation of the previous section applies only to a single isolated video object. For scenes with multiple objects, we must first decompose the scene into isolated objects and apply our formulation independently to each isolated object.

3. **Initial Shape Estimates**
   The containment, locality, and Voronoi Ordered Space components of E_{object} depend upon an initial shape estimate.

To address these three issues, we apply Principal Component Analysis to our multi-resolution optical flow to locate a conservative surface estimate. Our bootstrap stage finds regions of homogeneous motion that are used as the initial conservative surface estimate, $S_{initial}$ (see Figure 4.10).

PRINCIPAL COMPONENT MOTION FIELD SURFACE ANALYSIS

Given a motion field from our multi-resolution optical flow, we group pixels by their projected motion through time, treating the video sequence as a composition of *motion paths* (see Section 2.4). We can cluster these paths by their locality and projected motion through Principal Component Analysis (PCA) as shown in Figure 4.11. These clusters of homogeneous movement represent a conservative core regions of objects and bootstrap our surface optimization.

We use PCA to simplify our motion field [Kung et al., 1996] [Lin et al., 1996]. As mentioned in Section 2.4, we introduced a simple concept called *motion paths* that groups pixels by their correspondence of projected motion. Assuming pixels from the same objects are moving together, we can find the motion paths of different objects by the principal component of their projected motions. For example, for a 10 frame sequence, we derive a 20 length vector made up of the 10 x-y components of the projected motion for each motion path. After estimating the correlation matrix from the vectors of all motion paths, PCA extracts the first eigenvector from the correlation matrix of the dataset and the magnitude of the first eigenvector in each motion path. Next, we spatially cluster the PCA of our motion paths. Coupling our PCA with the initial path positions, the motion paths project into 3-D feature vector space. Since initial path positions are mutually exclusive, the projection of the motion paths into this feature space form a functional surface in 3-D feature space. After smoothing the surface, the peaks and saddles on this surface correspond to regions of homogeneous movement. Using estimated partial derivatives of this surface, we cluster these motion paths by their peak/saddle regions and derive initial conservative surface estimates.

BOOTSTRAP RESULTS

As shown in Figure 4.12, the bootstrap stage yields conservative estimates for initialization of our video object segmentation system. When our extraction of a motion field is accurate, the bootstrap stage finds core regions of objects by their homogeneous motion well and robustly. These regions become our conservative surface estimates, $S_{initial}$, and are suitable for surface initialization, multiple object separation and our containment, locality, and Voronoi Ordered Space components of E_{object}. However, since we use only motion in our object detection, we cannot find objects that move with the background.

The results from the MPEG-4 test sequences show the advantages and disadvantages of the bootstrap stage. For the coastguard sequence, we see that our bootstrap does its job quite well: the three foreground objects (large coastguard ship, smaller motorboat and the wake from the motorboat) are found. In the container sequence, the bootstrap stage detects the two objects (the small boat and the large container ship). Unfortunately, due to the occlusion of the white pole, the container ship is split into two distinct objects. In the current system, we cannot join these two objects back together, showing the need for higher-level anal-

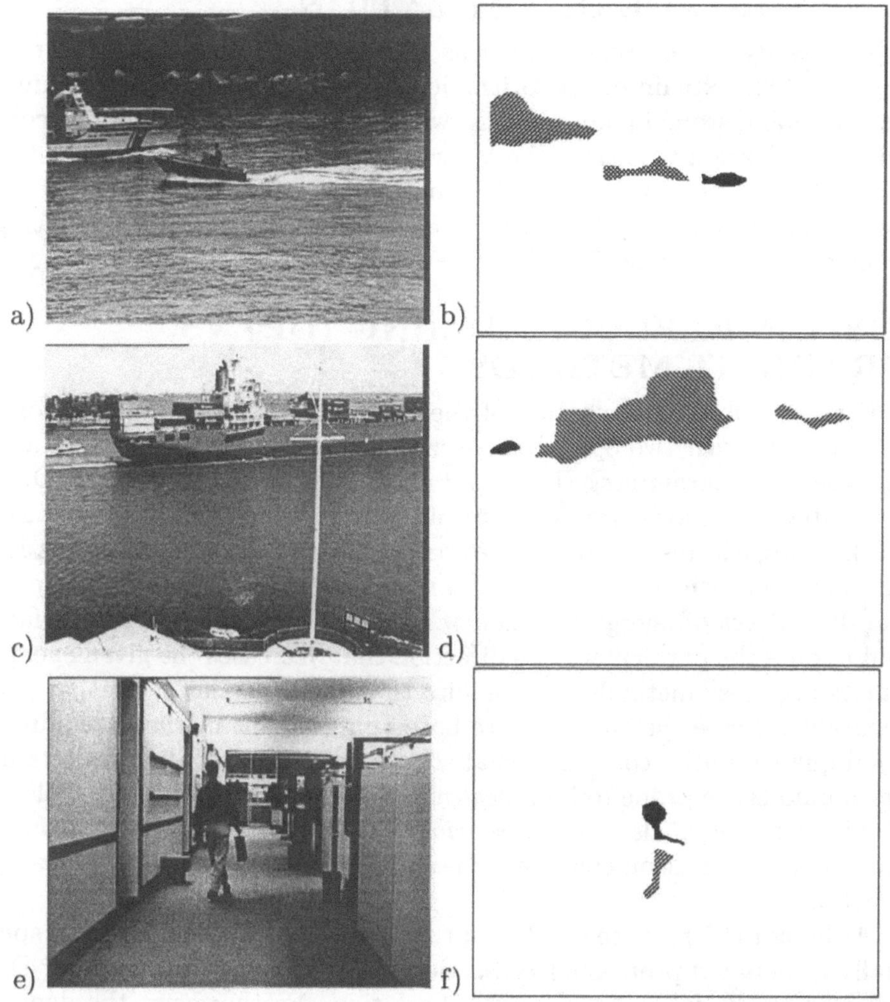

Figure 4.12. Bootstrap Results: for a) coastguard sequence, frames 30-39, b) container sequence 0-45, downsampled by 5 and c) hall monitor sequence, frames 30-39. Analysis of results are in section 4.12. Each differently textured region is a separate initial surface estimate in the frame.

ysis in our object detection. In the hall monitor sequence, the person is detected as two different objects, because of his non-rigid motion. These regions may also be joined together through some proximity measure, but, once again, our system currently has no high-level models of region interactions.

5. SURFACE OPTIMIZATION

We use dynamic programming as our surface optimization technology. We first explain our algorithmic choice of dynamic programming. As a stepping stone in our analysis, we show how we apply the 1-D contour optimization to extract the Viterbi Rings, the second component in E_{image}. We extend this dynamic programming algorithm for 1-D contour optimization to a 2-D surface optimization via iterative updating scheme, called Iterative Viterbi.

DYNAMIC PROGRAMMING (DP) VS. GRADIENT METHODS

For the global minimization of the surface energy function, we compare and contrast two optimization methods, 1) a gradient method and 2) dynamic programming (DP), and give our reasons for choosing DP.

Gradient methods are core optimization technologies in the snake tracking algorithms and neural network training. Gradient methods for optimization such as Newton-Raphson or Steepest Descent take the partial derivatives of energy function w.r.t. each parameter of the surface and update the parameters in a direction that decreases the given energy function. These methods only require that these partial derivatives are separable. These systems that are based on gradient methods require a good quality initial contour (usually from human input) because gradient methods are prone to find *local* minima in an optimization problem. Gradient methods also require *a priori* parameterization of the surface. This parameterization can affect the minimization of the surface energy function.

Although DP finds the global optimal solution, DP is limited to specially formulated problems (see Section 5.). The energy function for DP must be in the form of a recurrence relation. Furthermore, the dependency between subproblems induced by this recurrence relation must have an 1-D order, i.e., for all pairs of subproblems, $\vec{x}$ and $\vec{y}$, if $\vec{y}$ depends on $\vec{x}$, then $\exists f_{order}(\vec{x}) \rightarrow \Re$ s.t. $f_{order}(\vec{x}) < f_{order}(\vec{y})$. In this book, we use the following general form for the recurrence relation:

$$\max(E_{dynamic}(\vec{y})) = \max_{x \in \bar{\beta}(\vec{y})} (R(\vec{x}, \vec{y}) + c(\vec{x}, \vec{y}) \cdot \max(E_{dynamic}(\vec{x}))) \quad (4.13)$$

where we map our subproblems to points in the video space ($\vec{y}, \vec{x} \in \Re^3$), $\bar{\beta}$ returns the set of subproblems that are ordered before $\vec{y}$, i.e.,

$$\{\vec{z} | f_{order}(\vec{z}) < f_{order}(\vec{y})\}, \quad (4.14)$$

$R(\vec{x}, \vec{y})$ is the additive term for using a subproblem $\vec{x}$ to solve the subproblem $\vec{y}$, and $c(\vec{x}, \vec{y})$ are the weighted connectivities of positive value between subproblems. We find the optimal solution of $E_{dynamic}$ efficiently via the algorithm in Figure 4.13. Translation of our energy function $E(\mathcal{S}, I, M)$ into the compatible form of $E_{dynamic}$ may affect the search for our optimal solution of our desired energy function.

In our application, we choose a DP solution. DP can produce not only an optimized surface but also its parameterization [Amini et al., 1990]. The optimal solution from DP returns the set of optimal solutions of subproblems that were used in the calculation of the optimal solution and we can use this set to parameterize our surface. While we can dynamically change the surface parameterization in a gradient-based optimization by adding and removing parameters between the iterations, the DP algorithm gives us a more straightforward and elegant solution. For our system, the translation of surface energy function to a form acceptable to DP is not problematic.

In the future, we believe that a hybrid of dynamic programming and gradient methods would provide the best solution. DP can converge quickly on both a good parameterization of the surface and an good initial surface estimate. When surface parameterization becomes fixed, gradient methods integrate 2-D structure better and can use the original energy function. A hybrid method of surface optimization that alternates between the two methods of optimization is in our future research.

CONTOUR OPTIMIZATION BY DP: VITERBI RINGS

As a stepping stone in analysis of our surface optimization, we analyze a contour optimization algorithm that finds the Viterbi Rings with a circularly ordered space. The contour optimization, a variation of the Viterbi path-finding algorithm, is used within our surface optimization algorithm, but is applied over a Voronoi Ordered Space. DP requires three steps: 1) the formulation of energy function, 2) translation of the energy function into a recurrence relation that is compatible with DP, and 3) the application of DP to find the optimal contour. From this optimization, we then extract our Viterbi Rings.

We encode the heuristics of the Viterbi Rings as a contour optimization problem. The Viterbi Rings are a set of edges to be classified by probable relevance (to be defined shortly) to a given object center X. Edges are assumed to be more relevant when

DP-Path-Find $(f_{order}, \vec{x}_\circ, \vec{x}_{END}, R(\vec{x},\vec{y}), c(\vec{x},\vec{y}))$

```
{
  ARRAY mscore,backtrack;

  for x⃗ ∈ (β(f_order, x⃗∘, x⃗_END) ∪ {x⃗_END}) {
    mscore(x⃗) ← −∞ ;
    }
  mscore(x⃗∘) ← 0 ;
  for y⃗ ∈ (β(f_order, x⃗∘, x⃗_END) ∪ {x⃗_END}) in increasing f_order(y⃗) {
    for x⃗ ∈ β(f_order, x⃗0, y⃗){
      if(mscore(y⃗) < (R(x⃗,y⃗) + c(x⃗,y⃗) · mscore(x⃗)) {
        mscore(y⃗) ← (R(x⃗,y⃗) + c(x⃗,y⃗) · mscore(x⃗)) ;
        backtrack(y⃗) ← x⃗ ;
        }
      }
    }
  return backtrack ;
}
```

Figure 4.13. A general algorithm for dynamic programming. A particular problem can be specified the parameters: f_{order} is the 1-D ordering of subproblems, $\vec{x}_0$ is the given starting point, $\vec{x}_{END}$ is the ending problem, $R(\vec{x},\vec{y})$ is score contributions to maximize w.r.t. the path, $c(\vec{x},\vec{y})$ is the relationship between subproblems, and $\beta(f_{order},\vec{x},\vec{y}) = \{\vec{z} | f_{order}(\vec{x}) \leq f_{order}(\vec{z}) < f_{order}(\vec{y})\}$ returns the set of dependent subproblems that occur after $\vec{x}$ before $\vec{y}$ in the order. The algorithm returns the path in video space that maximizes the recurrence relation in Eq. 4.13.

1. edges are part of a closed contour whose sum total edge intensities are high

2. edges faces the object center X.

3. and edges are in a contour that circumnavigates the object center X.

For a given frame at time t', the first heuristic can be formulated as follows:

$$\arg_C \left(\max \left(\exp \left(\frac{\|C\| - L}{\sigma_C} \right)^2 \cdot \int_C R_{canny}(I\,(x,y,t)|_{t=t'}\,, \vec{s})d\vec{s} \right) \right) \tag{4.15}$$

where $I(x, y, t)|_{t=t'}$ is the frame at time t', C is the contour to be optimized, $\|C\|$ is the length of the contour, L is its expected length, σ_C is the variance of its length, and R_{canny} is the intensity map from the Canny Edge Filter applied to a frame of the video sequence. We restrict our search space of contours by describing the contour (C) in polar coordinates,

$$C = \left\{ \langle f(\theta)\cos\theta, f(\theta)\sin\theta \rangle \,|\, 0 \le \theta < 2\pi, \lim_{\epsilon \to 2\pi} f(\epsilon) = f(0) \right\} \tag{4.16}$$

where $f(\theta) \to \Re$ and θ is the angle of the point on the contour with the origin at the object center X. This parameterization encodes the second and third heuristics. Combining Eq. 4.15 and Eq. 4.16, we can define our search for Viterbi Rings as a search for a function that maximizes the given energy function for a contour estimate, $E_{ring}(C)$:

$$E_{ring}(C) = \exp\left(\frac{\|C\| - L}{\sigma_C}\right)^2 \cdot E'_{ring}(C) \tag{4.17}$$

$$E'_{ring}(C) = \int_0^{2\pi} R_{canny}\left(\left\langle \begin{array}{c} I(x, y, t)|_{t=t'}, \\ f(\theta)\cos\theta, \\ f(\theta)\sin\theta \end{array} \right\rangle \right) f(\theta) f'(\theta)\, d\theta \tag{4.18}$$

where C is the contour to be optimized, $\|C\|$ is the length of the contour, $f(\theta)$ is a description of C in polar coordinates, L is its expected length, σ_C is the variance of its expected length, and R_{canny} is the intensity map from the Canny Edge Filter applied to frame of the video sequence.

We translate E_{ring} in terms of the two functions of the recurrence relation of Eq. 4.13 that we call c_{ring} and R_{ring}, respectively. We define a subproblem $\vec{x}$ as the optimal score of a path that ends at a given point $\vec{x}$ in the video space, creating a one-to-one correspondence between subproblems and points on the isolated frame. To find the optimal path, we use the optimal paths that end at an adjacent point of lower θ value. The function $c_{ring}(\vec{x}, \vec{y})$ describes the relationship between subproblems: adjacency and the expected length of the contour:

$$c_{ring}(\vec{x}, \vec{y}) = \left\{ \begin{array}{cc} -\infty, & \vec{x} \text{ is NOT adjacent to } \vec{y} \\ \exp(\frac{\log(0.75)}{L}) & \textit{otherwise} \end{array} \right. \tag{4.19}$$

where $\vec{x}$ and $\vec{y}$ are subproblems/points in the same isolated frame and L is the expected length of the contour C. Since we wish to maximize edge intensity along the contour C on a isolated frame at time t',

$$R_{ring}(\vec{x}, \vec{y}) = R_{canny}(I(x, y, t)|_{t=t'}, \vec{x}) \tag{4.20}$$

where $\vec{x}$ and $\vec{y}$ are subproblems/points in the video space.

Using the algorithm in Figure 4.13, we can find good estimate of a contour (C) from the path returned from the following function call:

$$C = \text{DP-Path-Find}\begin{pmatrix} \theta(X, \vec{x}), \\ \vec{x}_\circ, \\ \vec{x}'_\circ, \\ R_{ring}(\vec{x}, \vec{y}), \\ c_{ring}(\vec{x}, \vec{y}) \end{pmatrix} \tag{4.21}$$

where X is the given object center, θ is a function that returns the counterclockwise angle from $\vec{x}$ point with X as the center, $\vec{x}_\circ$ is a starting point along the line $\theta(X, \vec{x}) = 0$, $\vec{x}'_\circ$ is the point along the line $\theta(\vec{x}) = 2\pi - \epsilon$ as $\epsilon \to 0$ that is adjacent to $\vec{x}_\circ$, R_{ring} is defined by Eq. 4.20, and c_{ring} is defined by Eq. 4.19. Since we know, at finite resolution, the number of points that are adjacent to a given point is bounded by a constant, this computation is $O(n)$ where n is the number of points. Pictorially, we can see that this DP is a variation of the Viterbi algorithm over a *circularly* warped space (see Figure 4.14), i.e., the polar coordinates. In our surface optimization, we apply the Viterbi algorithm over a Voronoi Ordered Space.

For Viterbi Rings, we find the set of contours that cover the edges that follow the heuristics. By iteratively finding the optimal contour, and removing intensity values from the $R_{ring}(\vec{x}, \vec{y})$ that coincide with the optimal contour, we find a concentric set of contours and this set of contours are our *Viterbi Rings* (see Figure 4.15a). This manageable and relevant subset of edges enables efficient use of our motion boundary coherence test (see Figure 4.15b).

ITERATIVE VITERBI

Iterative Viterbi applies the contour optimization techniques of the previous section within an iterative framework for surface optimization. DP requires a 1-D ordering/ parameterization of the surface: we must decompose the 2-D surface into a series of 1-D contours along the time axis. To maintain time smoothness of the surface while decomposing the problem, we expand our surface optimization over multiple iterations.

We decompose our problem into a set of 1-D problems that will be reintegrated together to form an estimate of 2-D surface. A dynamic programming solution requires that the subproblems of the original problem be ordered, i.e., parametrized and ordered w.r.t. a single variable. However, a straightforward decomposition does **NOT** approximate the surface optimization.

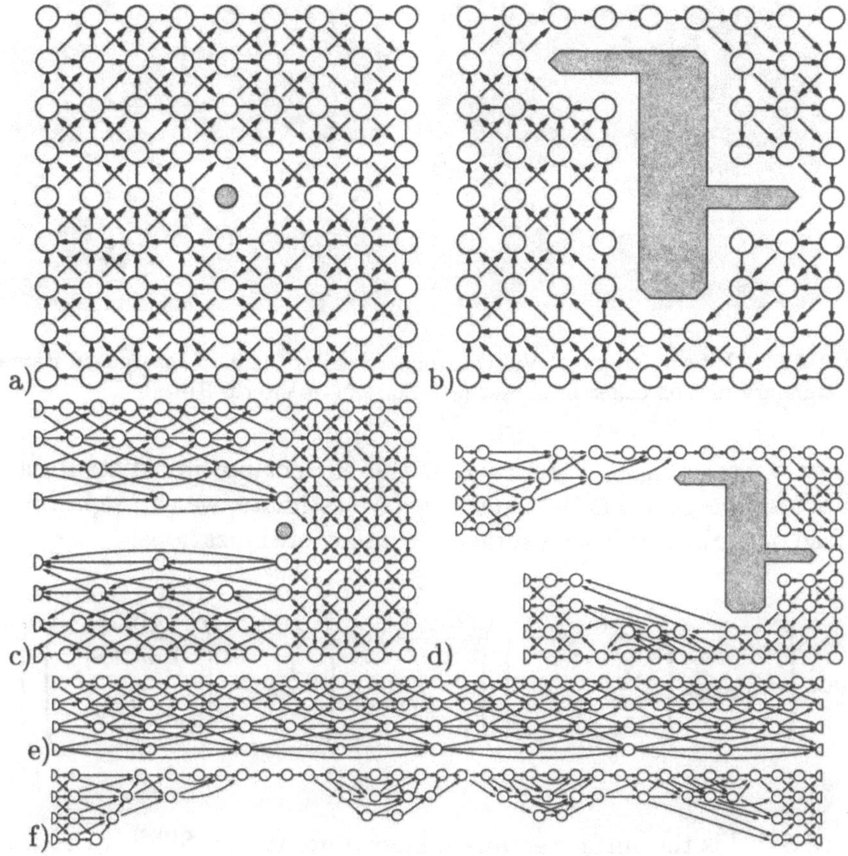

Figure 4.14. Viterbi Trellis, WarpedWarped Viterbi Trellis: the contour optimizations are equivalent to the Viterbi path-finding algorithm over a warped trellis diagram. a) we have our path calculations around a center (a circularly ordered space), c) the relationship between subproblems can be unwrapped around this center, e) the structure is equivalent to an irregular Viterbi trellis diagram. Similarly, b) we have our path calculations around an arbitrary contour (a Voronoi Ordered Space), d) the relationship between subproblems can be unwrapped around this contour, f) this structure is also equivalent to an irregular Viterbi trellis diagram.

$$\max_{\mathcal{S}} \left(E_{vobject} \begin{pmatrix} \mathcal{S}, \\ I, \\ M \end{pmatrix} \right) \not\approx \int_t \max_{\mathcal{S}^{(*,t)}} \left(E_{vobject} \begin{pmatrix} \mathcal{S}^{(*,t)}, \\ I, \\ M \end{pmatrix} \right) dt \tag{4.22}$$

If each contour is separately optimized, this decomposition destroys the time smoothness of the surface. For instance, two contours in adjacent frames may minimize their own energies individually, but the placement of each contour may cause the surface to be jagged, resulting in high

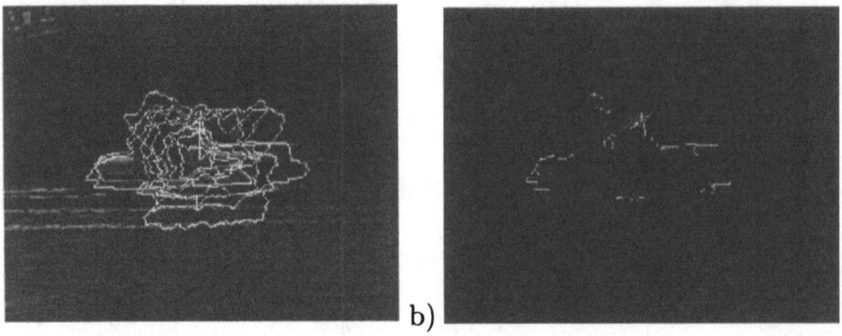

a) b)

Figure 4.15. **Viterbi Rings: a) Viterbi Rings (n=8) of the 93rd coastguard frame b) the boundary motion coherent subset (see Eq. 4.7) of Viterbi Rings**

energy w.r.t. E_{smooth}. However, if the current optimization iteration can access the previous iteration's surface estimate, we can separate the surface optimization into a series of contour optimizations:

$$\max_{\mathcal{S}^{(n,*)}} \left(E'_{vobject} \left(\begin{array}{c} \mathcal{S}^{(n,*)}, \\ \mathcal{S}^{(n-1,*)}, \\ I, \\ M \end{array} \right) \right) = \int_t \max_{\mathcal{S}^{(n,t)}} \left(E'_{vobject} \left(\begin{array}{c} \mathcal{S}^{(n,t)}, \\ \mathcal{S}^{(n-1,*)}, \\ I, \\ M \end{array} \right) \right) dt \tag{4.23}$$

where $\mathcal{S}^{(n,*)}$ is the surface estimate after n iterations, $\mathcal{S}^{(n,t)}$ is the timeslice at t of surface estimate after n iterations, and $E'_{vobject}$ is the modified energy function. All energy components in $E(\mathcal{S}, I, M)$ are separable w.r.t. time except for E_{smooth}. We can approximate time smoothness of the surface by modifying the smoothness component to include the surface estimate of the previous iteration:

$$E_{smooth} \approx E'_{smooth} \left(\mathcal{S}^{(n,*)}, \mathcal{S}^{(n-1,*)} \right) \tag{4.24}$$

We then may use DP to optimize a surface with a good approximation of time smoothness through iterative calculations:

$$\arg_{\mathcal{S}} \max E(S, I, M) \approx \lim_{n \to \infty} \mathcal{S}^{(n,*)} \tag{4.25}$$

In the next steps, we construct $c^{(n,t)}_{surface}(\vec{x}, \vec{y})$ and $R^{(n,t)}_{surface}(\vec{x}, \vec{y})$ for the DP recurrence relation, Eq. 4.13, that approximates $E_{vobject}$. A summary of the translation is given in Figure 4.16. To relate subproblems, we start with c_{ring} from Eq. 4.19 of the contour optimization and add locality and containment components.

Class of Analysis	Type of Analysis	Effect on Surface Optimization
Image	Edge Detection	Canny Edge Filter is the basis of $R_{surface}$
	Viterbi Rings	Only edges that are in the Viterbi Ring set undergo Boundary Coherence Test
Motion	Change Detection	All Edges values within changing regions are doubled in intensity.
	Motion Boundary Coherence	Intensities of edges that pass motion Boundary coherence tests and that are in Viterbi Rings are tripled
Environmental Assumptions	Smoothness	Sharp curvatures in the contours are disallowed; global search parameters are smoothed w.r.t. time; edges that fall between contour estimates of previous and next frames are strengthened as iterations continue; edges that are used in contour estimates are strengthened as well.
Object-Based Assumptions	Locality	Consider only contours that are some distance within w.r.t. the initial surface estimate.
	Containment	Consider only contours that contain the initial surface estimate.
	Voronoi Order	Consider only contours that are monotonically increasing w.r.t. the initial surface estimate.

Figure 4.16. The Translation of Energy Function into Dynamic Programming Parameters: this table summarizes how the energy components from Eq. 4.23 are translated into the Iterative Viterbi calculations Eqs. 4.26- 4.29

$$c^{(n,t)}_{vobject}(\vec{x},\vec{y}) = \begin{cases} -\infty, \vec{x} \text{ is NOT adjacent to } \vec{y} \text{ in frame t} \\ -\infty, \vec{y} \text{ is contained by } \mathcal{S}^{(0,t)} \\ -\infty, \vec{y} \text{ is further } \bar{d}(\mathcal{S}_{initial}) \text{ than from the } \mathcal{S}^{(0,t)} \\ \exp(\frac{\log(0.75)}{L^{(n,t)}}), otherwise \end{cases} \tag{4.26}$$

where $\mathcal{S}^{(0,t)}$ is a timeslice at time t of the initial surface, $L^{(n,t)}$ is the expected length of the contour $S^{(n,t)}$, $\mathcal{S}_{initial}$ is the initial surface estimate, and $\bar{d}$ is a distance measure based on the initial surface estimate. The Voronoi Order Space component will be added as the ordering of problems in Eq. 4.30. To construct $R^{(n,t)}_{vobject}(\vec{x},\vec{y})$, we combine our edge detection and change detection in step #1:

$$R_1^{(0,t')}(I,\vec{x}) = \begin{cases} 2R^{(0,t')}_{canny}(I(x,y,t)|_{t=t'},\vec{x}), & \vec{x} \in X_{change}(I) \\ R^{(0,t')}_{canny}(I(x,y,t)|_{t=t'},\vec{x}), & otherwise \end{cases} \tag{4.27}$$

where $I(x,y,t)|_{t=t'}$ is the isolated frame at time t', R_{canny} is the intensity map of Canny edge results for frame t', and X_{change} are the set of pixels that are marked as changing by our change detection analysis. We add our Viterbi Ring analysis in step #2:

$$R_2^{(0,t)}(I,\vec{x}) = \begin{cases} 3R_1^{(0,t)}(\vec{x}), & \vec{x} \in X_{bmc/vrings}(I) \\ R_1^{(0,t)}(\vec{x}), & otherwise \end{cases} \tag{4.28}$$

where $X_{bmc/vrings}$ are the subset of points in the Viterbi Rings that pass the test of boundary motion coherence. In our final step, we add our spatial and time smoothness components:

$$R^{(n,t)}_{surface}(\mathcal{S}^{(n-1,*)},\vec{x}) = \begin{cases} R_2^{(0,t)}(\vec{x}) + \delta(n) & \vec{x} \in X_{smooth}(\mathcal{S}^{(n-1,*)}) \\ R_2^{(0,t)}(\vec{x}) & otherwise \end{cases} \tag{4.29}$$

where X_{smooth} is the set of points from a smoothed version of $\mathcal{S}^{(n-1,*)}$, and $\delta(n)$ is a smoothness measure that increases with the number of iterations to converge the surface optimization.

The framework for the Iterative Viterbi Algorithm is shown in Figure 4.17. The decomposition of the problem into iterative framework divides the problem into two major stages, similar to Expectation and Maximization (EM) algorithm: an Optimization step and Re-estimation step. Applying the algorithm of Figure 4.13 within an iterative framework, we can calculate the surface through these steps:

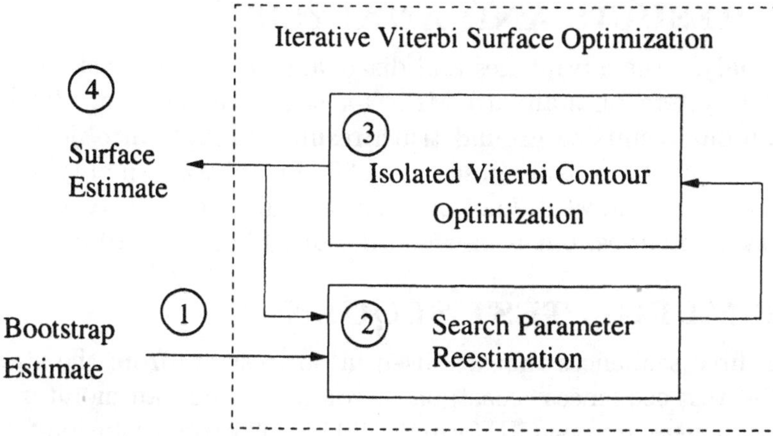

Figure 4.17. Iterative Viterbi Surface Optimization: 1. Initial Estimates are passed in from the Bootstrap Stage. 2. Each timeslice of the surface is optimized in isolation and form a surface estimate. 3. Search parameters are re-estimated based upon previous surface estimate. 4. The system iterates until the surface estimate converges.

1. Initialize ($n = 0$) surface estimate from bootstrap stage: $\mathcal{S}^{(0,*)} = \mathcal{S}_{initial}$

2. Estimate search parameters, $R^{(n,t)}_{surface}$,$c^{(n,t)}_{surface}$,$L^{(n,t)}$ and $\vec{x}_o^{(n,t)}$, from surface $\mathcal{S}^{(n,*)}$ and $\mathcal{S}^{(n-1,*)}$.

3. Individually optimize each contour of the surface at timeslice t of iteration n,

$$\mathcal{S}^{(n,t)} = \text{DP-Path-Find}\begin{pmatrix} \phi_V(\mathcal{S}^{(0,t)}, \vec{x}) \\ \vec{x}_o^{(n,t)}, \\ \vec{y}, \\ R^{(n,t)}_{vobject}, \\ c^{(n,t)}_{vobject} \end{pmatrix} \tag{4.30}$$

   where $\phi_V(\mathcal{S}^{(0,t)}, \vec{x})$ is the Voronoi Order from the tth timeslice of the initial surface estimate, $\vec{x}_o^{(n,t)}$ is an estimate of a starting point of minimum value of ϕ_V, $\vec{y}$ is a point adjacent to $\vec{x}_o^{(n,t)}$ of maximum value of ϕ_V such that it completes the closed contour, $R^{(n,t)}_{vobject}$ is from Eq. 4.29 and $c^{(n,t)}_{surface}$ is from Eq. 4.26.

4. Form surface $\mathcal{S}^{(n,*)}$. If $\mathcal{S}^{(n,*)} \not\approx \mathcal{S}^{(n-1,*)}$, then increment n by 1 and go back to step 2, else return the surface $\mathcal{S}^{(n,*)}$ as our video object estimate.

6. RESULTS AND ANALYSIS

We analyze the advantages and disadvantages for the system by running the system on standard MPEG-4 sequences [Sikora, 1997]. We compare our results to ground truth results through an objective segmentation quality measure in section 4.23. In our work, quality of results was paramount and we did not optimize for running time. Running time of the system varies, but is on the order of 10 hours for 10 frames.

THE MPEG-4 TEST SEQUENCES

The three sequences that are used in our test are from the standard MPEG-4 test sequences: coastguard, container and hall monitor. Each of the sequences are raw CIF format, 320 by 200 pixel resolution frame at 30 frames per second. The sequences are in color with YUV color space. Since the Y components approximate image intensity, we use only the Y channel in our calculations to approximate the grayscale images. We run our system on 10 frame subsequences of the video that conform to the simplifications in Section 2.. The results from these test sequences are shown as follows: Figure 4.18 are results from selected frames, Figures 4.20, 4.21 and 4.22 are complete results from the various test sequences, objective results are shown in Figure 4.24.

RESULTS FROM COASTGUARD SEQUENCE

The coastguard sequence (frames 30-39) has two foreground objects: the larger coastguard ship and a smaller motorboat with its wake. A short description of sequence follows: a large coastguard boat enters the screen from the left and moves right, a small motorboat moves to the right with water and shore in the background. Within this subsequence, the camera pans with the motorboat, while the coastguard boat moves to the right.

As shown in Figure 4.20, this sequence has particular qualities that are best suited for our algorithm. The foreground objects are rigid and their movements are orthogonal to the camera view, assuring that the object projection remains in the same size and shape except for its displacement. Thus, the temporal smoothness criterion works well in this case. Visually, the objects stand out from the background, allowing the edge detection algorithms to pick up the boundaries. The homogeneous background has a mixed effect on our calculations: it lessens noise on the edge detection, but the determination of background motion is more difficult. Our bootstrap stage finds the three objects in the video sequence. Since the optical flow is accurate, the bootstrap finds the objects in question, the large coastguard boat and the small motorboat,

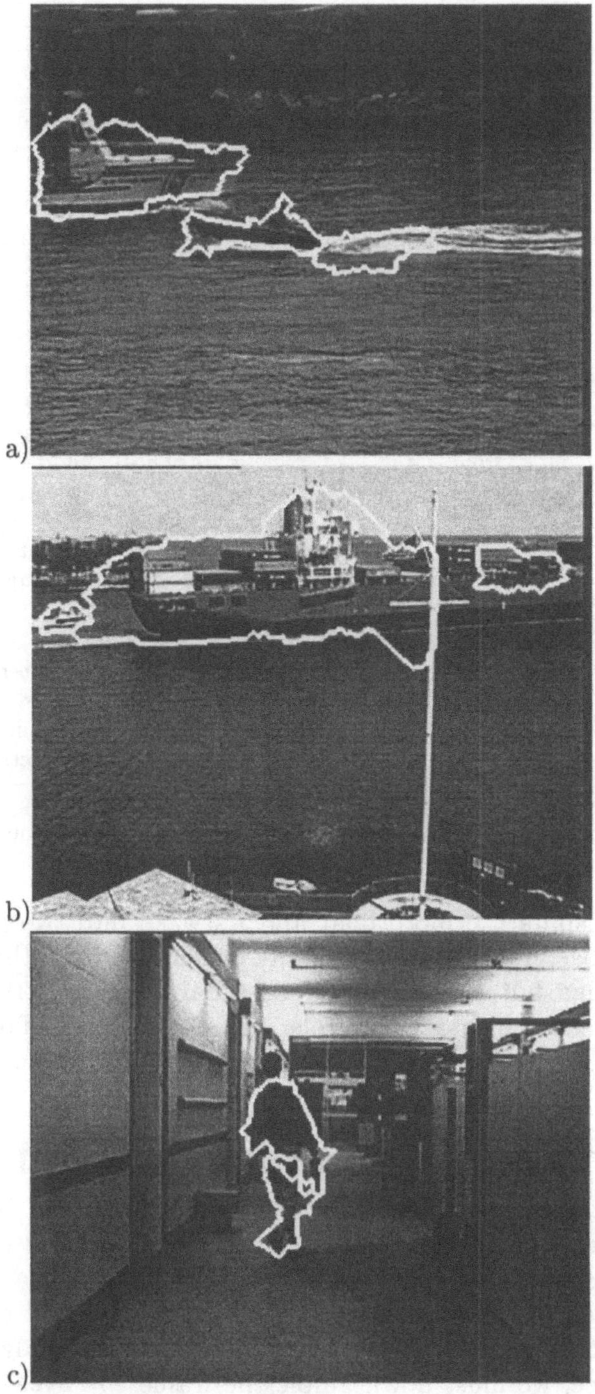

Figure 4.18. Selected results from MPEG-4 sequences; a)coastguard, frame 33 b) container (downsampled by 5), frame 15, and c) hall monitor, frame 63.

	Information Source		
Sequence	Image	Motion	Assumptions
Coastguard	Good - Edges are pretty well-defined and stand out from background, although there are a lot of internal edges.	Excellent - Objects are all rigid body; motion is orthogonal camera; different objects are moving in different directions; many texture regions for motion analysis.	Excellent - Objects are rigid bodies. Good estimates for Voronoi Ordered Spaces are available.
Container	Good - most edges stand out from the background	Poor - motion is slow to begin with; many key regions are of homogeneous intensity ; motion is parallel to many object boundary edges.	Poor - although object are rigid, in the middle of the largest object, occlusion occurs, splitting the container ship into two objects.
Hall Monitor	Good - most edges stand out from the background.	Good - Background is stationary so change detection is particularly accurate	Poor - Object is a non-rigid object and is composed of two differently moving sections.

Figure 4.19. Analysis of each of the MPEG-4 test sequences

although it splits the motorboat and its wake into two video objects. The large coastguard boat is partially occluded within the sequence by the frame edge, but when the large coastguard boat is more than halfway onto the screen, our system can tolerate some measure of occlusion. Objective results for the coastguard sequence (defined in Section 4.23) are shown in Figure 4.24.

RESULTS FROM CONTAINER SEQUENCE

The container sequence (frames 0-45, downsampled by 5) has two moving objects of interest: the large container ship and the small boat behind it. Since the objects are so far away from the camera, the projected motion of the objects between frames is beyond the pixel resolution and too small to be extracted by the motion analysis. Thus, for this sequence, we must downsample the frames by five. However, this manual preprocessing can be replaced by an automatic process based upon a simple activity measure.

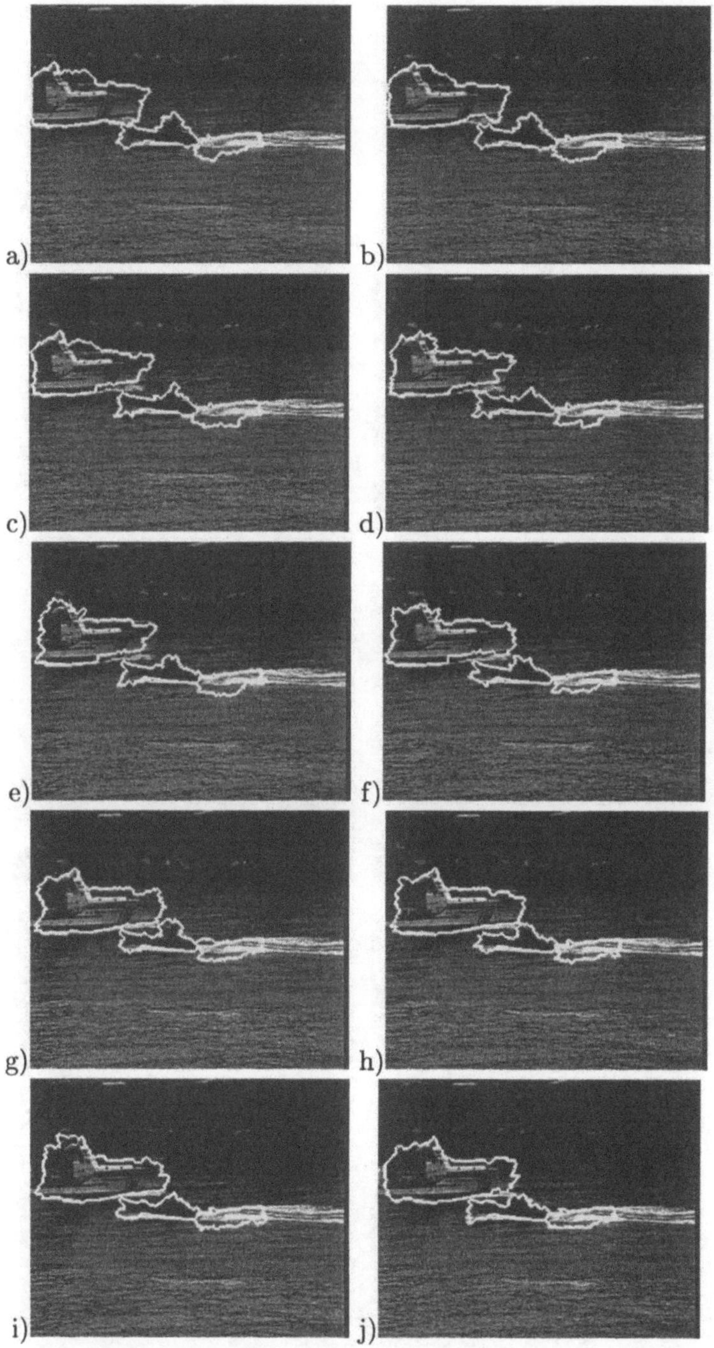

Figure 4.20. Segmentation Results from Coastguard sequence: Three objects were found, the large coastguard boat, the small motorboat and the wake of the small motorboat. Analysis of results are in section 4.19. $\vec{Q}$ results are shown in figure 4.24.

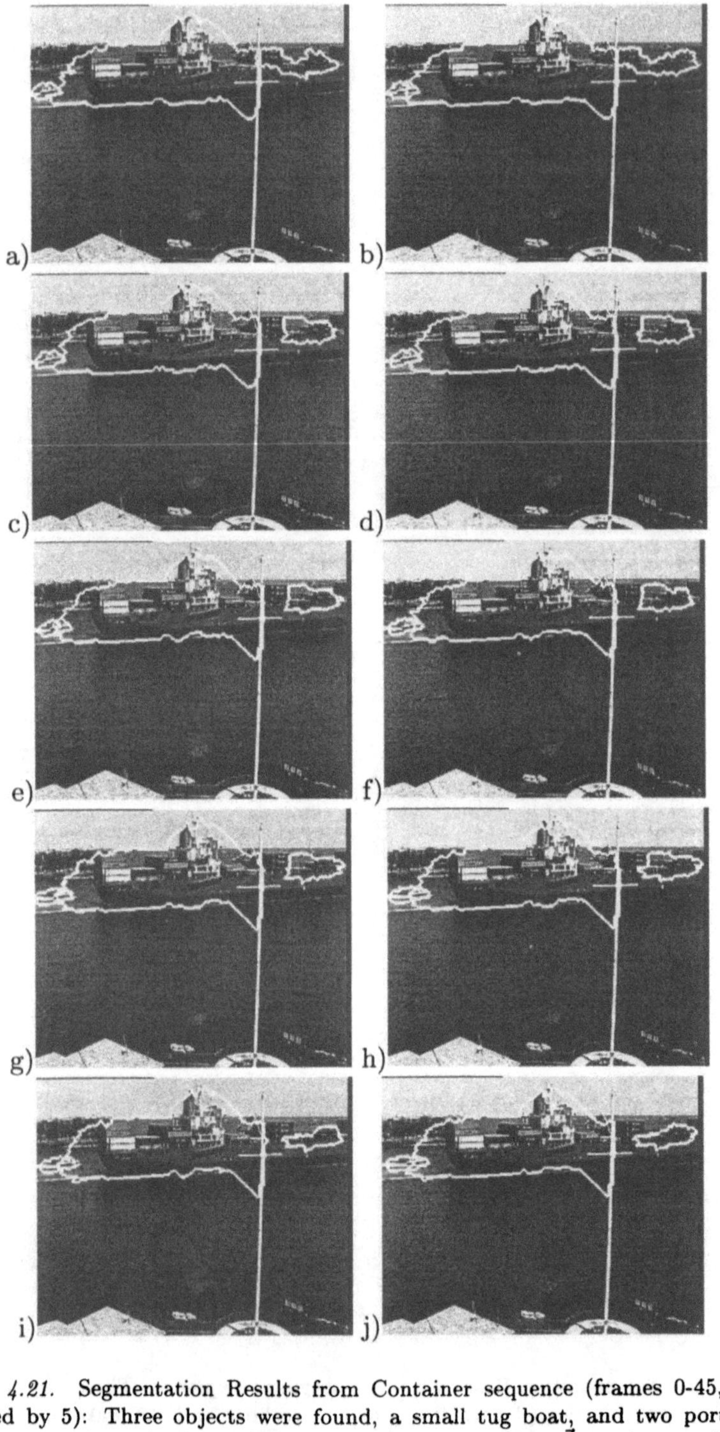

Figure 4.21. Segmentation Results from Container sequence (frames 0-45, down-sampled by 5): Three objects were found, a small tug boat, and two portions of the container ship. Analysis of results are in section 4.20. $\vec{Q}$ results are shown in figure 4.24.

As shown in Figure 4.21, this sequence is problematic because of difficulty of motion estimation. Most of the moving regions are homogeneous and confuse the motion estimation (an aperture problem). Thus, most information comes from the visual edges. Also, the white flagpole occludes the container ship and splits the initial object estimates into two objects. In the current system, this type of error is unrecoverable. In the future, high-level scene understanding may avoid these problems. Objective results (defined in Section 4.23) for the container sequence are shown in Figure 4.24.

RESULTS FROM HALL-MONITOR SEQUENCE

The hall monitor sequence (frames 60-69) has one person walking down the hall. The camera and background are stationary and the person is walking away from the camera.

As shown in Figure 4.22, this sequence is the most troublesome for our system. Since the figure is non-rigid, the bootstrap stage incorrectly finds two differently moving regions for the single object and, in the current system, we cannot recover from this problem. If we knew *a priori* that the background was stationary, then we could rely upon change detection more strongly. The adaptive or heuristic weighting of information sources are in our future research. Most regions on the person are homogeneous, giving little or no motion information. Unfortunately, there are no ground truth sequences for the hall monitor sequence, so no objective results could be calculated.

AN OBJECTIVE MEASURE OF SEGMENTATION QUALITY

In recent years, a major difficulty in evaluating results of content-based video processing is the lack of good objective measures. We introduce a simple measure called the 2-D quality vector, $\vec{Q}$, defined in Eq. 4.31 below:

$$\vec{Q} = \left\langle \frac{\|P_{truth} \cap P_{result}\|}{\|P_{truth}\|}, \frac{\|P_{truth} \cap P_{result}\|}{\|P_{result}\|} \right\rangle, \in [0,1] \times [0,1] \qquad (4.31)$$

where P_{truth} are the pixels in the ground truth segmentation and P_{result} are the pixels in the segmentation result. The components of the quality vector are called, respectively, the *content* and *coverage* percentage.

As shown in Figure 4.23, $\vec{Q}$ allows us to easily interpret the results from segmentation algorithms. An ideal segmentation algorithm would give results in $\vec{Q}$ of $< 1, 1 >$ and any refinement to the segmentation

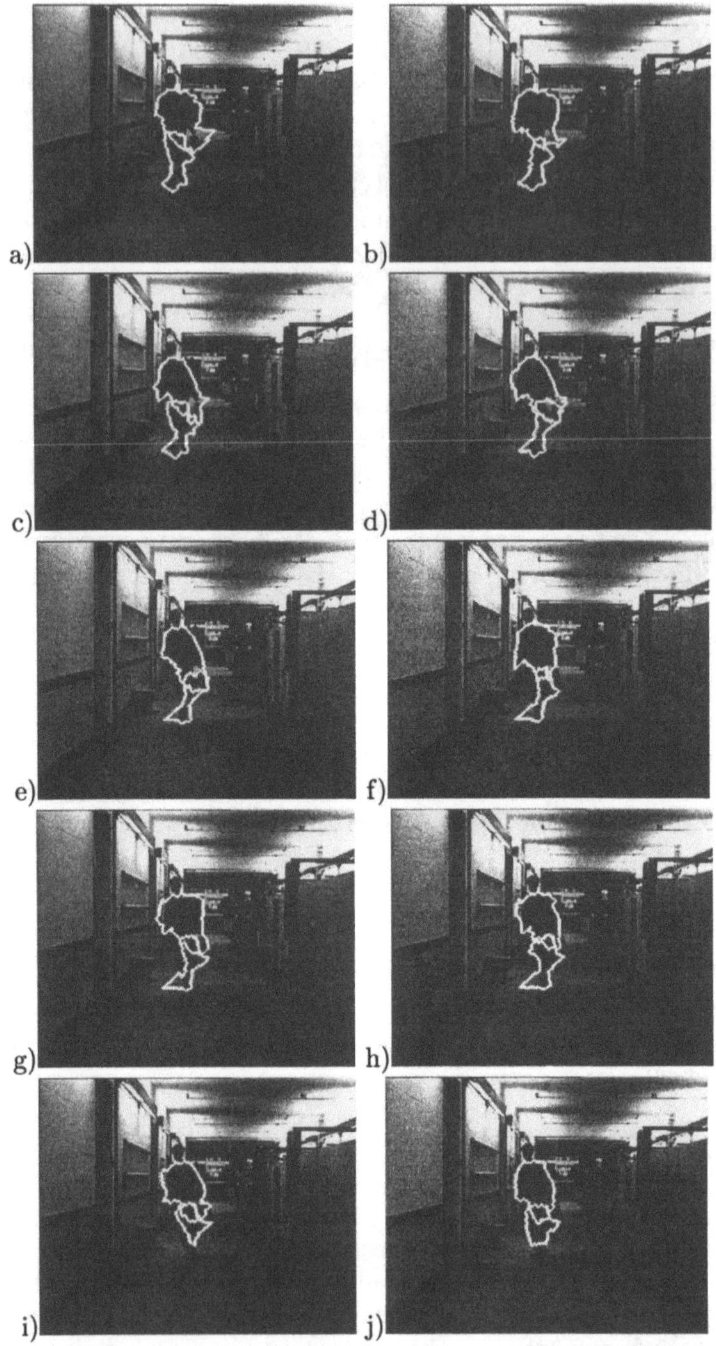

Figure 4.22. Segmentation Results for Hall monitor sequence (frames 60-69): Two objects were found that split the person into two sections. Analysis of results are in section 4.21.

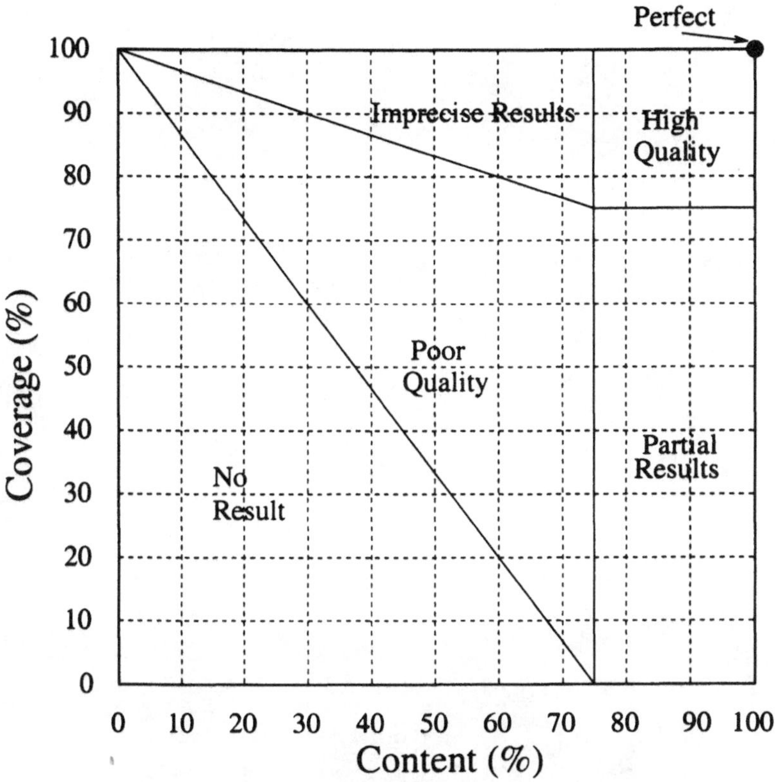

Figure 4.23. Qualitative Interpretation of $\vec{Q}$ (Eq. 4.31) in terms of segmentation quality

moves the $\vec{Q}$ toward $< 1, 1 >$. A high content, but low coverage result corresponds to the result of tracking system that finds only a small part of the object with high reliability. A low content, but a high coverage result corresponds to a result from a system that can localize area of the object, but cannot find the object within that area. The value of $\vec{Q}$ differentiates between bad results, good technique that could be further improved, and possible subjective bias in the ground truth results. The $\vec{Q}$ is similar to the concept of false positives and negatives in identification algorithms.

Our $\vec{Q}$ results of the coastguard and container sequence are shown in Figure 4.24. In the coastguard sequence, we see that we capture the large coastguard ship very well. In the ground truth segmentation, the motorboat and its wake are considered one object. Although the motor boat region is segmented reasonably well and shown to be a quality partial result, the segmentation of the motorboat wake is not especially good. In the container sequence, we see that we can get an imprecise

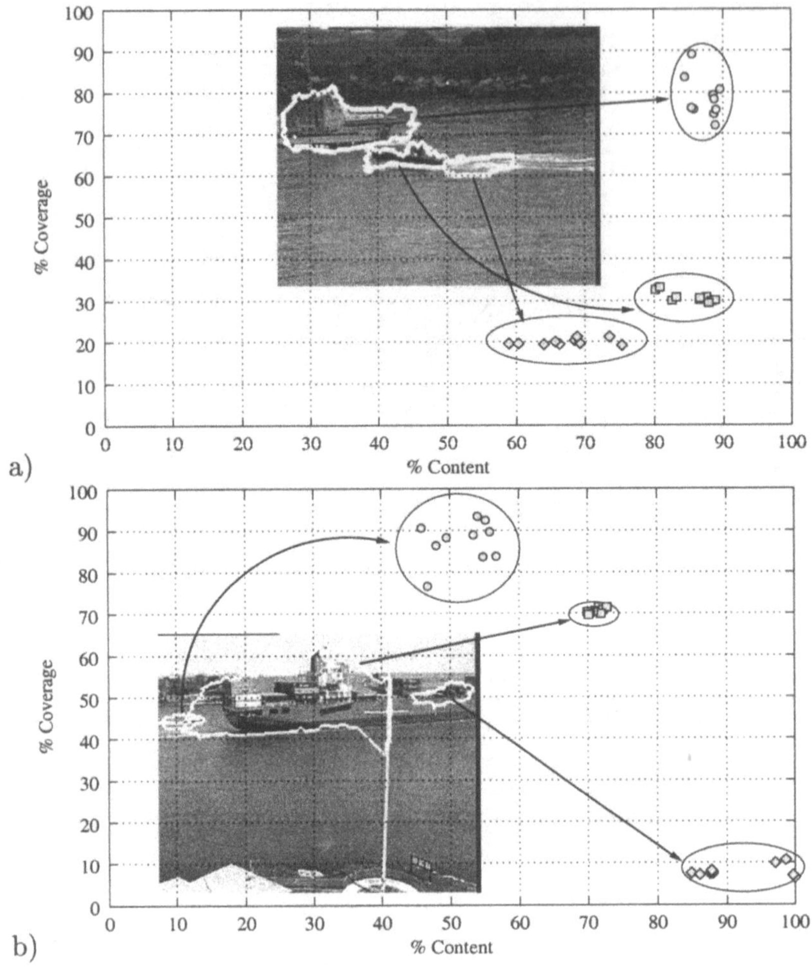

Figure 4.24. $\vec{Q}$ results of section 6. for coastguard (frames 30-39) and container sequence (frames 0-45, downsampled by 5). Each point represents the $\vec{Q}$ for each video object plane to the ground truth video object plane. The circles show the membership of the $\vec{Q}$ to the specific video object. Interpretation of results are in Section 4.23.

segmentation of the smaller boat. However, the white flagpole occludes the container ship and split it into two segmented objects: one of poor quality, and the other which can be considered a good quality partial result. Overall, the good clustering of points shows the good controllability of the surface optimization algorithm w.r.t. the time smoothing. Unfortunately, there does not exist ground truth data for the hall monitor sequence so we cannot do this type of analysis.

7. CONCLUSION

The MPEG-4 standard has established itself as the standard syntax for the first major step in content-based processing (e.g. cut and paste functionality, synthesis, query, etc.) [Committee, 1998]. The MPEG-4 standard supports the object-based coding of a video sequence, i.e., video objects, the partitioning of the video sequence into spatio-temporal regions that correspond to meaningful objects within the video sequence. This chapter implements the major enabling technology for this standard, a system for video object segmentation. We combine Voronoi Order and a surface optimization problem formulation for automatic video object segmentation. Voronoi Order describes a boundary search space given initial rough estimate of the object from optical flow. Our surface optimization formulation provides a fitness measure for object boundaries in the video sequence. By simultaneously maintaining an invariant Voronoi Order and maximizing the surface optimization metric, we extract the video objects. We analyze the results both qualitatively and objectively.

These extracted video objects are of marginal use unless users have some way of finding and browsing through them. In Chapter 5 and Chapter 6, we now turn to the problem of how to represent and search these video objects in order to allow users to find their desired video content.

Chapter 5

ROBUST REPRESENTATION OF SHAPE WITH DAGS

In this chapter, we review our general methodology of robust shape representation with Directed Acyclic Graphs (DAGs) [Lin and Kung, 1997a] [Lin and Kung, 1997b] [Lin and Kung, 1998b]. Chapter 2 outlined the techniques to extract relevant features of the video sequence, Chapter 3 showed the relationship between shape and extracted features, and Chapter 4 implemented a system to extract video objects themselves. To search the shapes of extracted video objects, this chapter lays the foundation of the concepts and algorithms for robust shape representation. Although the examples in this chapter come primarily from our work in on-line cursive handwriting recognition, the work can be generalized to drive our video object shape query system in Chapter 6.

1. PREVIOUS WORK

As mentioned in Section 1.9, the search and query functionalities for video objects proposed by the MPEG-7 standard are closely related to recognition technologies. In the context of the MPEG-7 standard, the encoding of video object shape is focused toward a representation that can be used to identify the class of content of a given video object. Since the relationship between the shape and the content class is a problem of not only shape representation, but also content recognition, many recognition technologies are applicable.

There are many schemes for solving recognition problems as vectors in a multidimensional feature space [Kung, 1993]. For data with structured complexity, there are also techniques such as time-dependent neural nets (TDNN), Hidden Markov Models (HMM) [Bellegarda et al., 1994] [W. et al., 1995] [M. et al., 1995] [Schenkel et al., 1994] [Garcia-Salicetti et al., 1995], curve representation techniques [Nishida, 1995]

How to use DAGs for Robust Recognition?

We give an example with on-line cursive handwriting.

Given a sequence of connected points that represent a letter, we can recognize a letter by seeing it as a series of loops and lines.

loop, line, loop line

However, as simple as process of converting a letter into a sequence of lines and loops, there is an inherent when we discretize the continuous line into a discrete sequence. The definition of our elements are naturally ambigous.

OR

loop,loop,loop,line

Our lowercase g may also have three loops.

OR

line,loop,loop,line

Or have two loops.

OR

line,line,loop, line

Or have one loop

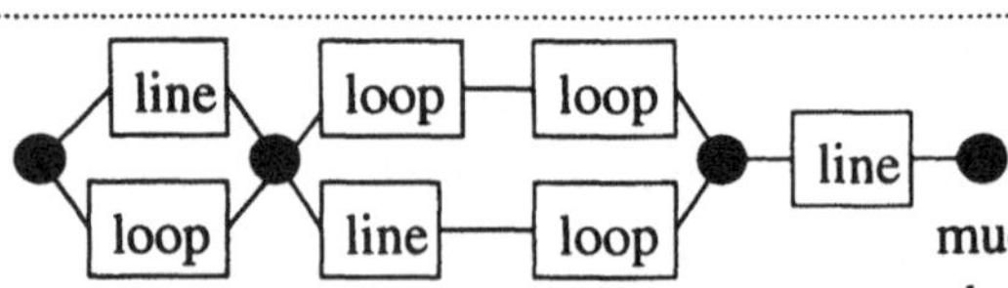

Since the four sequences share much of the same data, we can code all the sequences in one graph, a Directed Acyclic Graph (DAG).

Comparisons between these representations coded as DAGs implement a robust recognizer.

[Parizeau and Plamondon, 1995] [Singer and Tishby, 1994] [Schomaker, 1993] [Manke et al., 1994] and others that decompose the recognition problems to some extent [Cunn et al., 1997] [Man and Poon, 1993] [Fukushima and Imagawa, 1993] [Kadirkamanathan and Rayner, 1993]. In cursive handwriting and in speech recognition, HMMs are a popular and high quality solution [Rabiner, 1989]. However, the current HMM theory accepts only simple sequences. Our approach is fundamentally different in how we treat the problem. In traditional divide and conquer schemes for ordered information, a single larger problem is treated as a composition of more manageable subproblems. If this composition is mapped onto a graph, most methods treat the composition of the problem as a simple sequence. In contrast, we believe that the composition as a DAG is vital for a robust recognition. Finite State Transducers (FSTs) can accept these complex graph compositions [Mohri, 1997], but are less applicable to general signal processing because the current domain of FSTs is restricted to the discrete, finite-element languages. In recognition problems, FSTs generally applied to grammar modeling, word-lattices and other high-level constructs.

2. ORDER, MAPPING AND DAGS

Order aids in finding mapping between sets. When we use an order to compare sets, a data structure (such as a list, sequence or a DAG) that enforces an order over a set of data leads to efficient methods for determining a mapping.

Linear orders defined on two sets of data can be used to efficiently establish mapping between elements of the two sets by 1) restricting the set of possible mappings and 2) allowing divide and conquer techniques to be applied. An *ordered mapping* between sets (see Figure 5.1) is defined below:

$$a, b \in S_1, x, y \in S_2, if \left\{ \begin{array}{c} a \leftrightarrow x \\ b \leftrightarrow y \end{array} \right. then \left\{ \begin{array}{c} (a < b) \rightarrow (x < y) \\ (a > b) \rightarrow (x > y) \end{array} \right. \tag{5.1}$$

where S_1 and S_2 are the two ordered sets. This restriction drastically reduces the total number of possible mappings between sets. As shown in Figure 5.1c, if we are determining an ordered mapping, we can decompose the ordered mapping problem into two smaller independent ordered mapping problems by fixing a mapping between an element from each set. This type of decomposition allows a divide and conquer techniques to be applied.

These properties of linear order are also true for a *topological* order that exists in a DAG, i.e., an order of nodes such that every directed edge of the graph goes from a lower to higher ordered node [Cormen

What do we need to apply a DAG approach?

Certain properties are required to apply the DAG approach.

ORDER
- The DAG approach only applies to ordered information
- The information can be seen as a sequence
- With order, matching between these sequences becomes more efficient

PARTITIONING
- Ordered information can be grouped into a sequence of larger elements that can aid recognition
- However, this process may be cause ambiguities to be introduced into the representation
- DAGs will help us deal with these ambiguities.

DAG-CODING
- The ambiguities introduced by partitioning can be coded into the representation through the use of DAG
- Using a DAG-Coded representation we take advantage of the partitioning without losing information

DAG-COMPARISON
- In Partitioning and DAG-Coding, ambiguities are resolved by choosing the most similar structures between two DAGs and using this match as basis of comparison.

RECURSION
-The DAG approach is a generalized divide and conquer methodolgy, where partitioning and DAG-coding are a robust division and DAG-Comparison is the conquer part

-If a recognition problem can be divided at multiple resolutions, we can apply the DAG approach recursively at each level to implment to a our novel recognizer architecture.

If our objects have all these qualities, a DAG approach is a good fit for a recognition technologies.

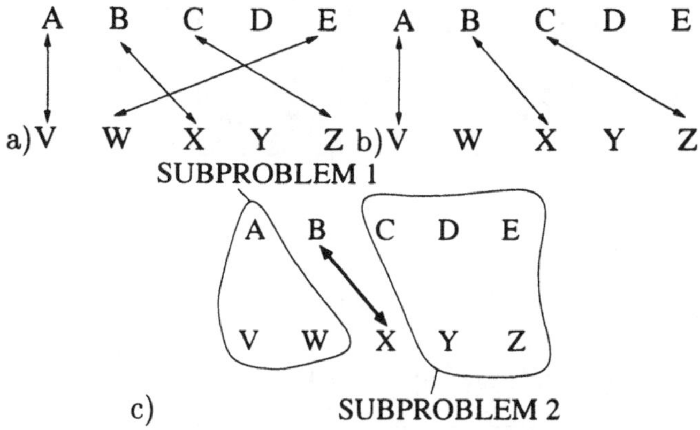

Figure 5.1. Order and Mappings: a) in general, mapping between sets (A-E) vs. (V-Z) can match any two letters from different sets b) if we enforce consistency of order (the alphabetical order) in our mappings as defined in Eq. 5.1, we disallow mappings between the two sets that cross each other c) the consistency allows for a divide and conquer approach for finding a mapping across two ordered sets.

et al., 1990]. As we shall see in section 5., although the topological order of a DAG is not necessarily unique, this consistency of the topological order allows us to apply divide and conquer methods to find a mapping between substructures of DAGs. By using DAGs as a robust representation, we can use the efficient calculation of an ordered mapping to robustly compare sets of data.

3. PARTITIONING

Our approach for recognition is based on the concept of *partitioning* (to be defined within this section). In speech and on-line cursive handwriting, the information is transmitted as *datastreams*, time-indexed series of real vectors. Recognition of a datastream as a whole is difficult unless we use order to decompose the datastream into smaller elements. This decomposition via partitioning is instrumental to the recognition process. Our discussion of partitioning in this chapter applies to only datastreams with 1-D order. In the next chapter, we extend these concepts to 2-D data.

We introduce two terms related to datastreams: the *macro-structure* and the *break*. *Macro-structures* are consecutive series of ordered vectors from the datastream that are related to only one sub-element of the datastream. For instance, in cursive handwriting word recognition, loops and lines are a consecutive series of points within a handwritten letter and can be considered macro-structures of the handwritten letter. In speech word recognition, syllables are macro-structures and

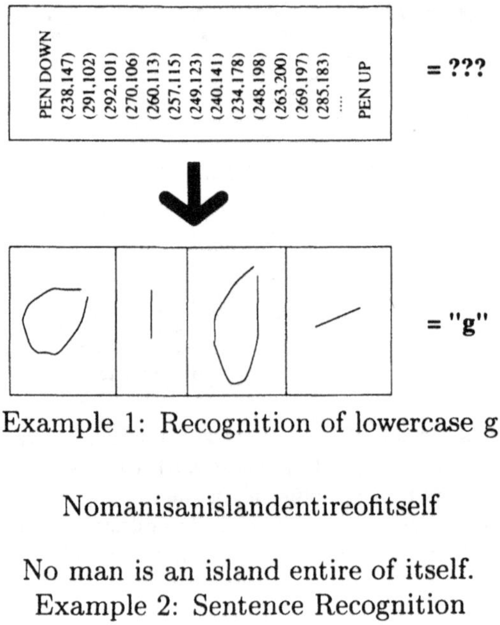

Figure 5.2. Partitioning to aid recognition: one example from cursive handwriting recognition; another from sentence recognition.

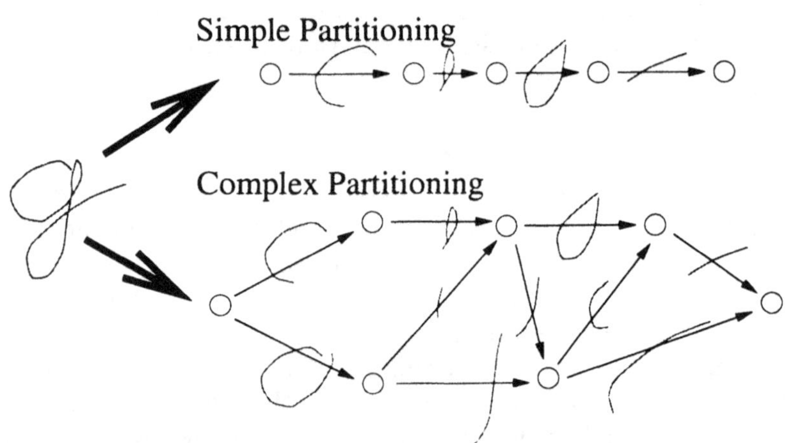

Figure 5.3. Simple vs. Complex Partitioning: shown in the top graph, the simple partition of sampled points can be represented as a list ; shown in the bottom graph, a complex partitioning contains a set of simple partitions and can be represented by a DAG.

can be formed into words. *Breaks* mark the boundaries of these macrostructures.

Partitioning is the process that identifies the breaks and induces the macro-structures from the position of the breaks. By finding the breaks in the datastream, we can distinguish this composition of macro-structures more readily than the datastream. As shown in Figure 5.2, if we can partition with only one macro-structure between two consecutive breaks, partitioning can aid in recognition by raising the level of representation of the datastream.

A *simple partitioning* inserts a series of breaks into the datastream and macro-structures are assumed to occur only between two consecutive breaks. A simple partitioning is the divide step in a traditional divide and conquer approach. A simple partitioning and its order may be expressed as a list of macro-structures. We generalize simple partitioning while maintaining the concept of ordering.

As shown in Figure 5.3, *complex partitioning* inserts a series of breaks into the datastream, but allows macro-structures to occur between any two breaks. If we are trying to recognize a unknown datastream, the location and existence of breaks may be uncertain and the definition of macro-structures may also be ambiguous. To be explained in Section 4., a complex partitioning can address these problems by expressing multiple simple partitions in a single data structure. To express complex partitioning in a data structure, we generalize the list that represents a simple partitioning to a DAG that represents a complex partitioning. This complex partitioning is the divide step in a *robust* divide and conquer approach that generalizes the divide step to complex partitioning.

4. DAG-CODING

DAG-Coding transforms a datastream to its DAG representation via complex partitioning. We explain how complex partitioning can be used for robust recognition, how complex partitioning maps to a specific type of DAG, DAG_o , and what characteristics of a datastream make DAG-Coded representation advantageous.

COMPLEX PARTITIONING FOR RECOGNITION

For recognition, complex partitioning supports a graphical representation that is robust to errors with the partitioning process and expresses some of the inherent complexity of the datastream. Partitioning locates breaks within the datastream without prior knowledge of the macro-structure. Ideally, a simple partitioning can locate all breaks in a datastream, but, either due to 1) the errors in the break location heuris-

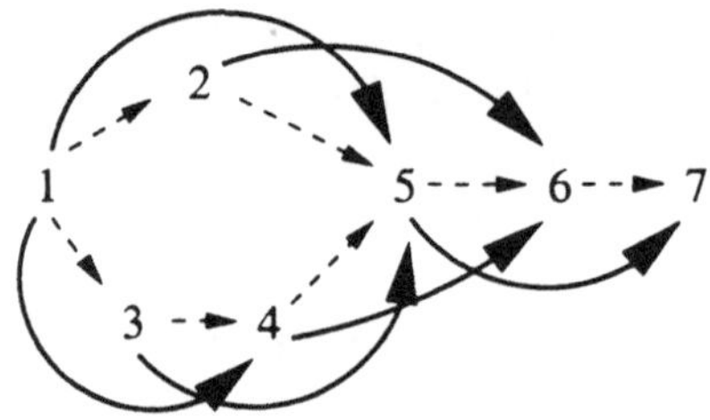

Figure 5.4. A Complex Partitioning that deals with imperfect partitioning. Spurs are the breaks that are wrongly inserted into the datastream. With additional edges to bypass any number of a non-consecutive spurs, this complex partitioning can account for these spurs. The original edges are dashed; the solid, dark edges are added to account for a non-consecutive spurs.

tics or 2) the ambiguity of the macro-structure definition, a complex partitioning is usually required.

Complex partitioning can compensate for imperfect partitioning. Partitioning attempts to break the input stream into pieces that are assumed to contain a single macro-structure, but it is a heuristic process. Simple partitioning assumes all breaks are correct, but, in a recognition problem, this requirement for correctness is too stringent. In complex partitioning, we can code our representation in such a way that it is tolerant to the heuristic-based break insertion methods that are only generally correct. Breaks that are wrongly inserted into the datastream are called *spurs*. Complex partitioning deals with non-consecutive spurs by adding edges to represent macro-structures that bypass these spurs (See Figure 5.4), yielding representation that is robust to spurs.

When the boundaries of macro-structures overlap or are uncertain, complex partitioning can represent this ambiguity so that it may resolved at a higher level of context. Even if break locations are found correctly, we can create ambiguity in our representation by inducing macro-structures and raising the level of representation of the datastream (e.g., in speech, from syllables to words). For instance, in Figure 5.5, we have two examples: the coding of a datastream of points as a complex partitioning of lines/loops and the coding of a datastream of syllables as a complex partitioning of words. In a simple partitioning, a part of the datastream belongs to only one macro-structure and we would have to immediately resolve any ambiguity without the benefit of context. In a complex partitioning, a part of the datastream may belong to multiple macro-structures and we can express ambiguity through this multiple membership.

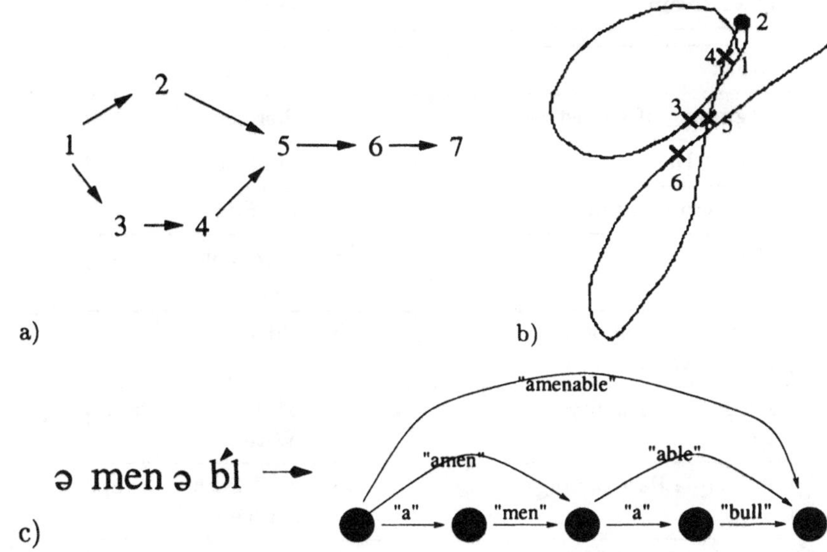

Figure 5.5. a) The DAG structure for b) a cursive lowercase G, partitioned at loop and cusp points. The multiple paths represent the loop/cusp duality. c) the transformation of the spoken syllable into a word-level DAG raises the level of representation. In cursive handwriting letter recognition, macro-structures are loops and lines; breaks in cursive handwriting are pen-lifts, pen-downs, cusps and loop crossings. In speech sentence recognition, macro-structures in speech are words; breaks in speech are pauses or frequency changes.

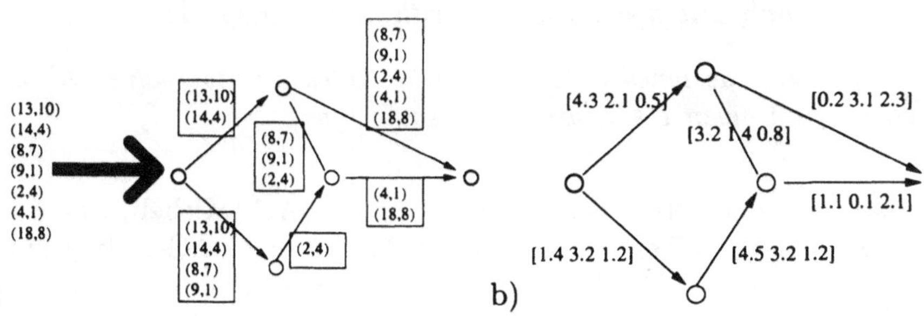

Figure 5.6. Example of a Continuous Stream of points converted to a DAG with edge values as three dimensional vector. a) a datastream after partitioning and b) DAG-Coded datastream after partitioning and extracting edge values.

MAPPING A COMPLEX PARTITIONING ONTO A DAG

We define $DAG_\circ$ as a special subset of DAGs and the mapping of a complex partitioning onto a $DAG_\circ$.

Datastream	$\rightsquigarrow$	$DAG_\circ$
Break	$\Leftrightarrow$	A Node
Beginning of Datastream	$\Leftrightarrow$	Source Node
End of Datastream	$\Leftrightarrow$	Sink Node
A Macro-structure	$\Leftrightarrow$	An Edge
Order	$\Leftrightarrow$	Directionality of Edges
Representation of a macro-structure	$\Leftrightarrow$	Edge Value
Ambiguity, Duality	$\Leftrightarrow$	Multiple Path With $DAG_\circ$
A Simple Partitioning	$\Leftrightarrow$	A Path from source to sink
A Set of Simple partitions	$\Leftrightarrow$	$DAG_\circ$

Figure 5.7. Correspondence between $DAG_\circ$s and a datastream

DEFINITION 5.1 *A $DAG_\circ$ is a DAG with the following restrictions:*

1. *There is only one node, the source, with no incoming edges.*
2. *There is only one node, the sink, with no outgoing edges.*
3. *Each edge is assigned a data structure called its edge value. All edge values in a given $DAG_\circ$ are of the same type.*

Our basic data structure is similar to a polar DAG [Micheli, 1994], but with *edge values.* When we refer to a $DAG_\circ$ in our DAG-coding applications, we assume a DAG with the restrictions in Definition 5.1. For a specific example of a simple DAG coding, refer to Figure 5.6.

Macro-structures and their implied relations between the ordered breaks create a representation with a $DAG_\circ$ in which every path from source to sink represents a simple partitioning of the input datastream, i.e., a partitioning with a linear order. A full correspondence between a $DAG_\circ$ and a datastream is given in Figure 5.7. In a $DAG_\circ$, breaks are represented by nodes, with the source as the start of the datastream and the sink as the end of the datastream. The information contained between two breaks is represented by an edge. Edge values of the graph represent the datastream between two breaks under the assumption that only one macro-structure exists between the two breaks. A path from source to sink corresponds to a sequence of macro-structures that is equivalent to a simple partitioning. A $DAG_\circ$ expresses a set of simple partitions.

For a robust recognition process, we wish to compare all simple partitionings of two datastreams against each other. By using $DAG_\circ$ as a compact expression of this set of simple partitionings, we can efficiently compare two complex partitionings and the sets of simple partitionings that they encode.

APPLICABILITY OF DAG-CODING

Although any datastream can be DAG-Coded, this section lists the characteristics that a datastream must have to benefit from representation in DAGs:

1. **Existence of Macro-Structures**
   The partitioning process attempts to break up the datastream with respect to macro-structures. If there are no such macro-structures within the datastream, partitioning has no meaning.

2. **Determination of Break Locations without Datastream Content**
   We assume we can locate the breaks without knowledge the datastream content. As in most content-based segmentation problems, we cannot truly identify the content without correct partitioning, and we cannot partition the datastream without knowing the content. To break this cyclic dependence, we require the position of breaks can be inferred directly from the datastream with good accuracy.

3. **1-D Ordering within the Datastream**
   DAGs require the datastream to be ordered. This ordering allows a path within the DAG to compactly represent a simple partition of a datastream. If there is no ordering of the macro-structures, the paths within DAGs no longer correspond to simple partitionings, invalidating the representation. Most datastreams have this ordering enforced through time indexing, although any 1-D ordering of the data is satisfactory.

4. **Partitioning Granularity**
   Efficiency and quality of the partitioning is directly related the reduction of data granularity in its extraction. If the partitioning is too fine, DAG-Coding will have little advantage over the original datastream, since the number and ordering of elements are similar. If the

partitioning is too coarse, conventional feature extraction methods through single functional mapping are a simpler solution than DAG-Coding. In general, DAG-Coding needs to balance order with the complexity of edge value.

5. **Cost of Redundant Information**
   The complex partitioning tends to represent the same part of the datastream on different edge values. The price of redundancy is hopefully compensated by improved recognition performance.

5. COMPARING DAG REPRESENTATIONS

This section presents an efficient algorithm (DAG-Compare) to measure similarity between two DAG-coded representations. The DAG-Compare algorithm can be used for classification by comparing DAG-Coded an unknown input to DAG-Coded examples that have previously classified. Given two DAG-Coded representations, the DAG-Compare operation finds the optimal match (to be defined in the next section) between any simple partition represented in one DAG and any other simple partition represented in another DAG. In terms of the graph, this operation finds the optimal match between any path in one $DAG_\circ$ and any path in another $DAG_\circ$. After deriving a sequence of edge values (the representations of the macro-structures) for each path, we can describe a path matching as a Longest Common Subsequence problem [Cormen et al., 1990] on these two sequences. For those readers familiar with the dynamic programming, this operation is computationally equivalent to finding a shortest path route through the crossproduct of two $DAG_\circ$ s that have had null self-loops added [Ramadge and Wonham, 1989] [Tsitsiklis, 1989] and this section is optional. However, since this operation is integral to understand the applications of the DAG-coded representation, we provide the complete analysis to aid those less familiar with dynamic programming.

COMPARING PATHS

If we describe a path as a sequence of edge values, we match two sequences, A and B, through a function η called PathScore, defined below:

$$\eta(A_m, B_n) = \max \begin{pmatrix} \text{MatchEdge}\left(\lambda, b_n, \eta(A_{(m)}, B_{(n-1)})\right), \\ \text{MatchEdge}\left(a_m, \lambda, \eta(A_{(m-1)}, B_{(n)})\right), \\ \text{MatchEdge}\left(a_m, b_n, \eta(A_{(m-1)}, B_{(n-1)})\right) \end{pmatrix} \quad (5.2)$$

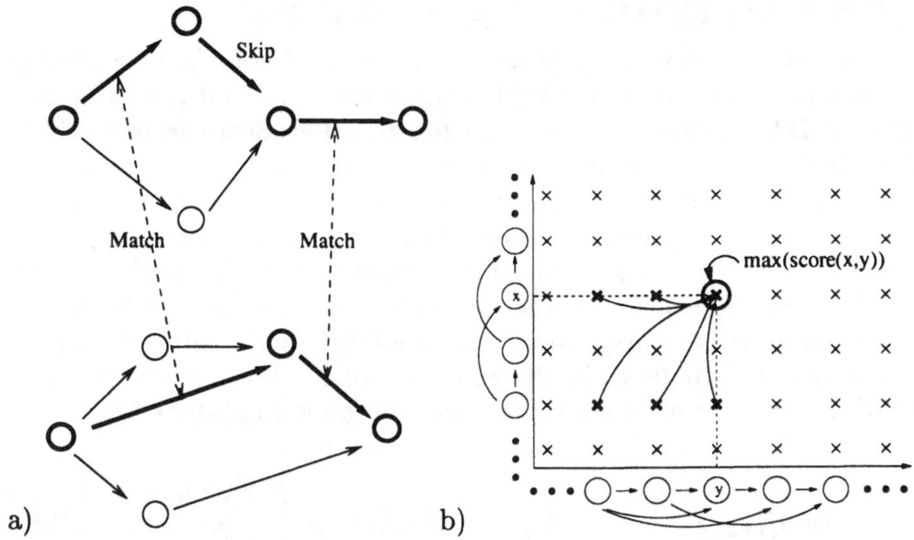

Figure 5.8. a) PathScore calculation: node and edge correspondence within the PathScore definition: Darkened nodes and edges corresponds to the two paths that give the highest matching score. The dotted lines show which two edges were matched together. b) Computational Dependencies for dynamic programming algorithm for path comparison of two DAGs: The previous problems are represented by the crosses. Note that all previous problem triangles are contained within the shaded area.

where a_m is the mth edge value in sequence A, b_n is the nth edge value in sequence B, $A_m = (a_1, \ldots, a_m)$ is subsequence of edge values from the first to the mth from sequence A, and $B_n = (b_1, \ldots, b_n)$ is the subsequence of edge values from the first to the nth from sequence B, λ is a null edge, and the MatchEdge function relates the previous comparison score of two subsequences to the next score after integrating the comparison of an edge from each sequence (see Figure 5.8). The formulation of MatchEdge is further restricted to be monotonically decreasing with respect to the previous score, i.e.,

$$\forall a, b : (x < y) \rightarrow MatchEdge(a, b, x) < MatchEdge(a, b, y). \tag{5.3}$$

The selection of MatchEdge is up to designer, but should model the matching probability and informational loss associated with comparing the two edge values as the representations of macro-structures. A better match between edge values can mean less of a decrease in score and a perfect match can avoid any loss in score.

THE DAG-COMPARE ALGORITHM

This section presents a polynomial-time algorithm for path matching between two DAG_os called DAG-Compare where we find the two paths for each DAG_os that maximize the function PathScore from Eq. 5.2. The matching score between two paths can be solved in quadratic time [Cormen et al., 1990]. All pairs of paths between two DAG may be compared via dynamic programming (DP) in quadratic time.

For DP, we identify subproblems associated with this problem: the function score(v_1,v_2) returns the set of all path scores for one path that goes from source and to a node v_1 in one DAG_o and another that goes from source to a node v_2 in the another DAG_o . We can describe this set of scores in terms of the recurrence relation with MatchEdge.

$$score(v_1, v_2) = \bigcup_{\substack{e_1 \in InEdges(v_1), \\ e_2 \in InEdges(v_2), \\ c \in score(\text{Prev}(e_1), \text{Prev}(e_2))}} \{MatchEdge \begin{pmatrix} V(e_1), \\ V(e_2), \\ c \end{pmatrix}\} \tag{5.4}$$

where $InEdges(v)$ are the set of edges going into node v, the function V returns the edge value of a given edge, and the function Prev returns node that an edge leaves. The DAG-Compare algorithm finds $\max(score(s_1, s_2))$ where s_1 and s_2 are the sinks of two DAGs. The solution to the current subproblem can be reached by traversing at most one edge in each graph from a previous subproblem. To find maximum of this set, this formulation still leads to a possibly exponential time computation. By using MatchEdge's monotonicity, we can simplify the equation for the maximum of a given score set:

$$\max(score(v_1, v_2)) = \max \left(\bigcup_{\substack{e_1 \in InEdges(v_1), \\ e_2 \in InEdges(v_2),}} \{MatchEdge \begin{pmatrix} V(e_1), \\ V(e_2), \\ c(e_1, e_2) \end{pmatrix}\} \right) \tag{5.5}$$

$$\text{where } c(e_1, e_2) = \max(score(\text{Prev}(e_1), \text{Prev}(e_2)))$$

By the dynamic programming principle, we recognize that only the maximum score needs to be remembered at any previous step.

There exists an order of solving the maximum of the score sets that preserves these edge dependencies, i.e., all computation of previous scores

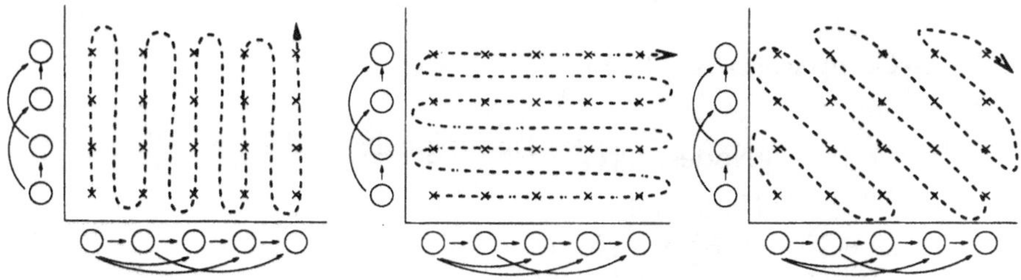

Figure 5.9. Valid Orders of Computation

can be finished before traversing another pair of edges. In fact, there are many orders of solving the subproblems that preserve these dependencies (See Figure 5.8b and 5.9). Since both graphs are DAGs, we can topologically sort [Cormen et al., 1990] the nodes in each graph such that all edges leave from one node and lead to another of higher order. Thus, a subproblem is only dependent on other subproblems that are contained within the rectangular area between the problem and the origin. Graphically speaking, the inductive process of solving subproblems can be pictorially viewed as growing an area along the axes such that any point to be added to the area is supported both horizontally and vertically from each axis.

If we calculate the subproblems in an order as described in the previous paragraph, we find the maximum PathScore of all paths encoded between two DAGs with MatchEdge as defined in Eq. 5.2. To find the correspondence between edges of the two DAGs actual paths, we keep track of which optimal solutions were used in optimal final solution and follow its predecessors from the final solution at to the initial solution. The pseudocode for this algorithm is shown in Figure 5.10. If MatchEdge is an O(1) operation, the running time of our algorithm is:

$$O((|V_1| + |E_1|)(|V_2| + |E_2|)) \tag{5.6}$$

where (V_1, E_1) and (V_2, E_2) are the sets of nodes and edges of the two $\text{DAG}_\circ$s to be compared.

APPLICATION: CURSIVE HANDWRITING LETTER RECOGNITION

In our own work, we have applied complex partitioning and DAG-Coded representations to the recognition of on-line cursive handwriting letter recognition [Lin and Kung, 1997a]. Using loops and curved line segments as our macro-structures, we implemented a cursive handwriting letter recognizer.

```
DAG-Compare(DAGo R,T)
{
   (r_1, r_2, ..., r_||Nodes(R)||) ← TopologicalSort(R);
   (t_1, t_2, ..., t_||Nodes(T)||) ← TopologicalSort(T);
   for (r ∈ Nodes(R), t ∈ Nodes(T)) {
       δ[r, t] ← −∞;  # Initialize scores
       p[r, t] ← −∞;  # Initialize scores
   }
   δ[r_1, t_1] ← 1.0;  # Set b basis of induction at (source,source)
   for i ← 1 to ||Nodes(R)|| {
     for j ← 1 to ||Nodes(T)|| {
       for e_r ∈ InEdges(r_i) {
         for e_t ∈ InEdges(t_j) {
             a ← Prev(e_r);
             b ← Prev(e_t);
             if(δ[r_i, t_j] < MatchEdge(e_r, e_t, δ[a, b])) {
               δ[r_i, t_j] ← MatchEdge(e_r, e_t, δ[a, b]);
               p[r_i, t_j] ← (a, b);
               }
           }
         }
       }
   }
   return δ[r_||Nodes(R)||, t_||Nodes(T)||],p ;
}
```

Figure 5.10. DAG-Compare: an algorithm for Path Matching between two DAGs. The function Prev returns the node that given edge leaving, InEdges returns the set of edges that coming into a given node, Nodes returns the nodes associated with a DAG_o , the function TopologicalSort returns an topologically ordered list of nodes from a given DAG and MatchEdge is the function defined in Section 5..

6. RECURSIVE STRUCTURE OF DAGS

The DAGs and the DAG-compare operation are a divide and conquer method that creates a recursive system architecture. When used recursively to extract the edge value in a DAG-Coding, a DAG-Coder and its DAG-Compare operation can be considered as the inductive step within a recognition system architecture. Although DAG-Coded representation varies at each level of recursion, DAG_os and their DAG-Compare oper-

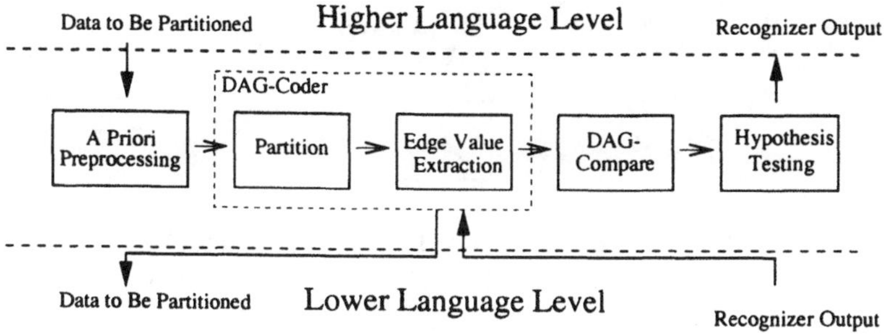

Figure 5.11. Inductive Step for DAG-Based Recursive System Architecture: 1. A Priori processing is defined as all processing of the information that can be done without knowledge of recognition results. 2. Partitioning As mentioned in Section 4., the next step is to divide the problem into smaller sub-problems while still maintaining the relationships between sub-problems. 3. Edge Value Extraction is when the DAG structure of the information is created, the values of the edges must be computed. 4. DAG-Comparison 5. Hypothesis testing is any post-processing information that uses the recognizer hypothesis output within its computation.

ation can be repeatedly applied to successive sub-problems within the same system. A system can recursively apply complex partitioning to divide the recognition problem into recognition of its macro-structures and uses the DAG data structure to encode and maintain relationships among macro-structures. Going into the recursion, the system then recurses on the macro-structures and divides them into smaller macro-structures until they are simple enough to be handled by a simple recognizer without partitioning. Coming out of the recursive structure of the system, the recognizer repeatedly applies the DAG-Compare operation on the recognized macro-structures to integrate contextual information and to recognize the composition of macro-structures. In Chapter 6, we will combine this recursion with our DAG_o data structure to form a new data structure for 2-D shape representation.

INDUCTIVE STEP

As shown in Figure 5.11, each system level that uses DAG_o and the DAG-Compare is an inductive step in a recursion that has two major purposes: 1) to decompose the problem into a set of macro-structure recognition problems and 2) to recognize the datastream through a composition of macro-structures. The first purpose is accomplished through partitioning and recursion; the second through the DAG-Compare operation. Complex partitioning divides the problem in a robust manner, tolerant to partitioning errors. The recursion on each edge effectively encapsulates each call on the recognizer at the lower level, allowing in-

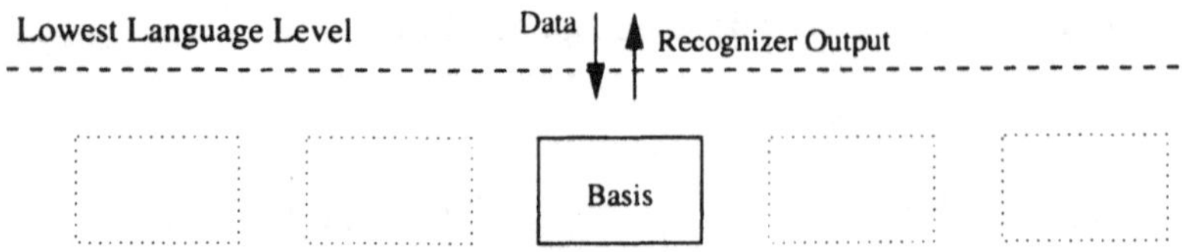

Figure 5.12. Basis for DAG-Based Recursive System Architecture

dependent computation of each edge value. When the recursion returns, its output is used in the edge value representation of the macro-structure. After the DAG structure is derived and all the edge values have been computed, the DAG is compared against a reference DAG with the DAG-Compare operation. A DAG-Compare can perform any operation that can be expressed as a path comparison between two DAG$_{\circ}$s such as grammar checks or dictionary searches.

BASIS

As shown in Figure 5.12, the basis of these systems is a special case of the inductive step where the datastream does not require partitioning to be recognized. When the datastream cannot be decomposed any further, a single functional block handles the recognition of the datastream. This functional block can be complex and/or adaptive, such as a HMM, TDNN, Kohonen self-organizing feature map or as simple as the low frequency points of a DFT. The classification results from this functional block are passed back up. A recursive design philosophy would suggest a basis to be as simple as possible.

APPLICATION: CURSIVE HANDWRITING WORD RECOGNITION

We have implemented a simple cursive word recognition system [Lin and Kung, 1997b] that has a simple design with two levels of recursive structure with a simple Discrete Fourier Transform (DFT) curve recognizer as lowest level (or basis) system. As shown in Figure 5.13, the recursive structure of the cursive word recognizer integrates a curve matching algorithm, letter recognizer and a full dictionary search. In cursive handwriting, breaks exist at different levels of representation: loop/line, character, word, and sentence level. Each level of representation corresponds to a level within the recursive architecture. This structurally recursive architecture and its implications on recognizer design are discussed in detail in [Lin and Kung, 1997a].

Level	Grammar	Word	Letter
Node	Word Breaks	Letter Breaks	Loop/Line Breaks
Edge	Possible Words	Possible Letters	Curve Segment
Edge Values	A Vector of Words with confidence scores, quantized meaning, tenses, types and cases	A Vector of Letters confidence Scores	DFT Vector with relative X,Y position
Reference DAG	Grammar Graph Structure	Dictionary Trie	Preclassified DAG-Coded Letter Examples
Functionality	Grammar Verification	Full Dictionary Search	Letter Recognizer

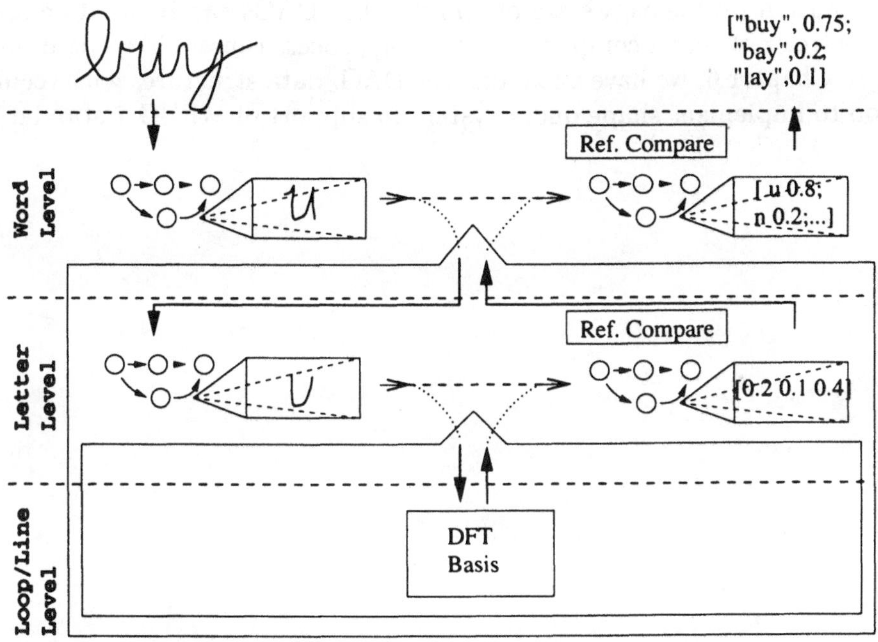

Figure 5.13. A Recursive architecture for On-line Cursive Word Recognition that uses DAGs [Lin and Kung, 1997b]. First, the word is broken down into letters via first level of DAG-Coding. The letters are broken down into loops and lines via a second level of DAG-Coding. The letters and lines are represented through a Discrete Fourier Transform and used as edge values on the second level of DAG-Coding. Once all the edge values are available, then a DAG-Comparison is applied so that the letter recognition results can be passed back to the first level. With the letter recognition results, the first level applies a DAG-Comparison to find the word recognition results.

7. CONCLUSION

This chapter applies Directed Acyclic Graphs (DAGs) to a large class of pattern recognition problems and other recognition problems where the data has a linear ordering. The datastreams are DAG-coded for robust partitioning. The similarity of two datastreams can be manifested as the path matching score of the two corresponding DAG_os . This chapter also presents the DAG-Compare algorithm an efficient and robust dynamic programming algorithm for comparisons of two DAG_os. Since the DAG-Coding provides a robust partitioning process, it can be applied recursively to create a novel system architecture. Not discussed in this chapter, the DAG structure also allows adaptive restructuring, leading to a novel approach to neural information processing [Lin and Kung, 1998b]. DAG-Coding may also be applied to any datastream where a complex partitioning aids the recognition process. Although the majority of the initial work was done on a test bed of handwriting recognition problems, we are of opinion that DAGs can be used on any datastream where a complex partitioning process can aid in recognition.

In Chapter 6, we have expanded the DAG_o data structures with recursion to implement shape query system in support of MPEG-7 standard.

Chapter 6

A SYSTEM FOR IMAGE/VIDEO OBJECT QUERY BY SHAPE

In this chapter, we combine our DAG Representation from Chapter 5 and our concept of Voronoi Ordered Space from Chapter 3 to find video objects in a database that are similar to a given shape. Within the context of the MPEG-7 standard, tools such as Image/Video Query by Shape (shape query, for short) fulfill an important functionality toward the creation of a "new" media (see Figure 6.1): a search over terabytes of Internet video to find relevant video content. After receiving the desired video content, users can manipulate, blend and arrange this content with their own content to produce new multimedia documents.

1. PREVIOUS WORK

In this section, we place our work in relation to image query. We review shape representation methods and link our work to a previous body of work on Voronoi Skeletons that leverage assumptions about the object content to model shape deformations [Ogniewicz and Kubler, 1995].

While we consider only the shape characteristics of a single objects in this chapter, the image query problem searches images that can be composed of many different objects. Image query systems [The and Chin, 1988] [Menon et al., 1996] [Hafner et al., 1995] [Lin et al., 1997] [Smith and Chang, 1997] [Remias et al., 1996] [Yu and Wolf, 1997] depend upon the composition of a scene, rather than the properties of a single video object. We assume that the video objects are available (e.g., from an MPEG-4 stream) while the image query systems must work on unsegmented images. In this way, we differentiate the image query problem from shape query problem. While image query blends much segmentation work into its analysis, we treat video object extraction and representation separately.

Some shape query methods treat the shape as a signal with no assumptions about the content either as a 2-D area or as a 1-D contour. Area-based methods such as moment-based shape descriptions [Prokop and Reeves, 1992] [Khotanzad and Hong, 1990] [Bailey and Srinath, 1996] treat shape as the signal over a 2-D space while contour-based method such as a Generalized Hough Transform [Jeng and Tsai, 1991] [Samal and Edwards, 1997] and curvature-scale space [F. and A.K, 1992] treat the shape as signal over 1-D order. However, even the choice of an area-based vs. contour-based representation biases the concept of similarity. For instance, thresholding has very different effects dependent upon the shape representation. A long, but thin protrusion may be removed by thresholding in an area-based representation while it may be an important feature in a contour-based representation. While shape seems to be an objective feature of a video object, its interpretation depends strongly on the method of representation.

If we have some *a priori* knowledge about a class of objects, we can leverage this knowledge into the design of our shape query system. If we know how an object shape deforms, we can add invariance to these deformations into our shape query system. For instance, a template-based shape matching allows full flexibility along the range of its parameters [Fischler and Elschlager, 1973]. Our shape extraction is based upon previous concepts of Euclidean distance, shape thinning algorithms, and a concept of Voronoi Skeletons [Cantoni and Carrioli, 1981] [Mayya and Rajan, 1995] [Ogniewicz and Kubler, 1995]. The representation is a novel recursive data structure based on work in Chapter 5 that allows us to robustly represent these Voronoi Order Skeletons (VOS). Shape query systems that use *a priori* knowledge can only be applied reliably to a subset of object shapes: in our case, shapes that contain an physical skeletal structure. However, as long as we recognize the underlying assumptions of our shape descriptor, we effectively apply this shape descriptor to search for desired video content.

2. PROBLEM DESCRIPTION

Although the evaluation of system results for shape query is subjective, we provide a general shape query problem description, shape query terms and guidelines on how to interpret results.

The problem of shape query is described as follows. In the context of video object search of the MPEG-7 standard [Group, 1999], given a query shape as a black and white image, find the shapes within the database that are manually classified as the same class of content. The

Why do we use Shape for Content Query?

There are many reasons why we focus on shape:

QUERY BY SHAPE WILL BE A COMMON QUERY

As video becomes more accessible, video database cannot rely on manual annotation, so video object shape is a way to infer content automatically: shape is an intuitive non-verbal way of describing an object content.

SHAPE IS SUPPORTED BY MPEG-4

Within the MPEG-4 framework, the basis of shape descriptor is encoded into every MPEG-4 video object. A MPEG-4 video database has all the requisite, raw shape information, ready for query technologies.

SHAPE IS AN EXPRESSIVE QUERY TOOL

Shape contains much object information about classification, pose and activity. Furthermore, the video object shape also evolves with time, allowing the evolution of a object shape as a basis of queries for particular actions (such as running or waving).

SHAPE IS ONE OF MANY QUERY TOOLS

Shape alone will not define a user query, of course. Shape query is a means to filter video content . For a single query, many different content filters may be required and multiple iterations of different content filters will be used in a single query. Content filters for color, size and movement, working together with shape can provide a powerful realization for content searches.

Content Query by Shape is one of many tools that provide an algorithmic infrastructure for consumer navigation of video content database

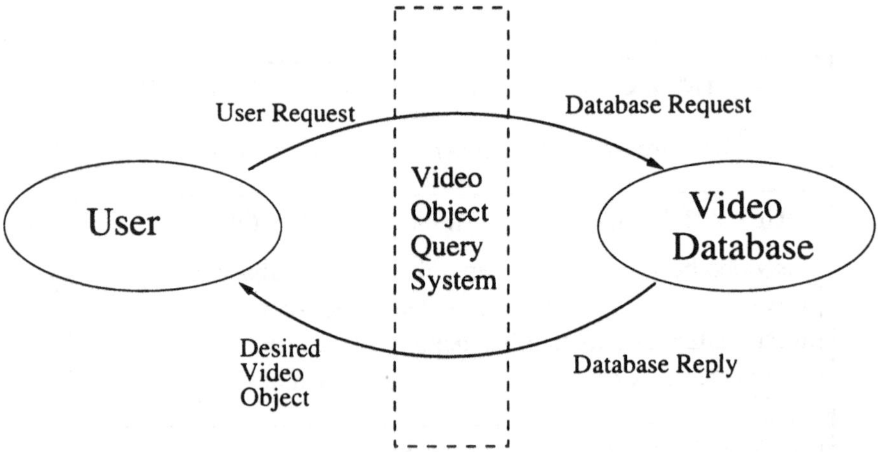

Figure 6.1. Image/Video Object Query: a transparent interface layer between users and Video Databases, translating users requests into database actions and database references into intelligible results.

representation for a shape is called a *shape descriptor*. The shape query system implements a *similarity measure* that, given two shapes, returns a value between 0 and 1, a higher value for two shapes that are more similar, via comparison of shape descriptors. With this query shape, the shape query system returns the n shapes that score the n-highest similarity measures. The user may then refine the search by changing the query and repeating the search. Although the evaluation of a similarity measure is affected by assumptions made about the object class (see Figure 6.2), we can list desirable qualitative properties of shape descriptor and its similarity measure:

1. **Accuracy**
   An accurate similarity measure scores the shapes in the same content class as high and those in different content classes as low. Precision is secondary: in MPEG-7, we wish to search for classes of objects, rather than one particular object.

2. **Robustness**
   Ideally, similarity measures of objects in the same content class should be invariant to noise, object rotation and object scaling. Since these shapes exist in a video environment, the similarity measure of objects in the same content class should also be invariant to different viewing angles and object movement.

3. **Controllability**
   User-aided query systems can leverage user feedback to find more

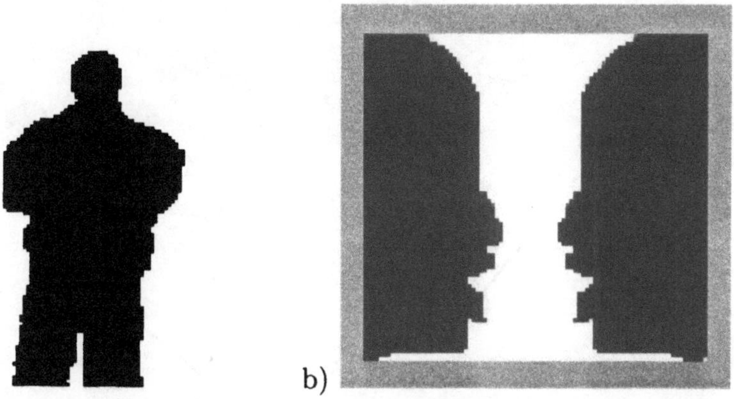

a) b)

Figure 6.2. The shape of object projection can aid classification and provide content. For instance, in a), if we classify the object as a person, we can infer the height and pose of the person. Shape is also naturally ambiguous and the same shape have very different meanings such as b), where the picture shape can either be a two faces or a candlestick.

easily and more reliably desired content than an automatic query system. A user interface that allows manual guidance of similarity measure can aid the accuracy of the query system.

4. **Speed**
   Computational cost of calculating the similarity measure should not be a limiting factor: thousands of shapes should be compared in a few minutes. The computational cost of extracting the shape descriptors is a secondary concern, since the extraction is done off-line.

5. **Compactness**
   Shape descriptors should be on the size of the compressed raw bitmap or smaller.

For our shape descriptor in particular, we also describe a new functionality:

- **Motion Annotation**
  A descriptor that captures the temporal aspect of video information provides users with more powerful and descriptive search criteria. Since the shapes exist within the context of video sequence, motion annotation of shape descriptors may be a powerful query tool.

3. SHAPE EXTRACTION VIA VOS

In this section, we describe how we extract shape information by using assumptions about object content and derive a representation that is

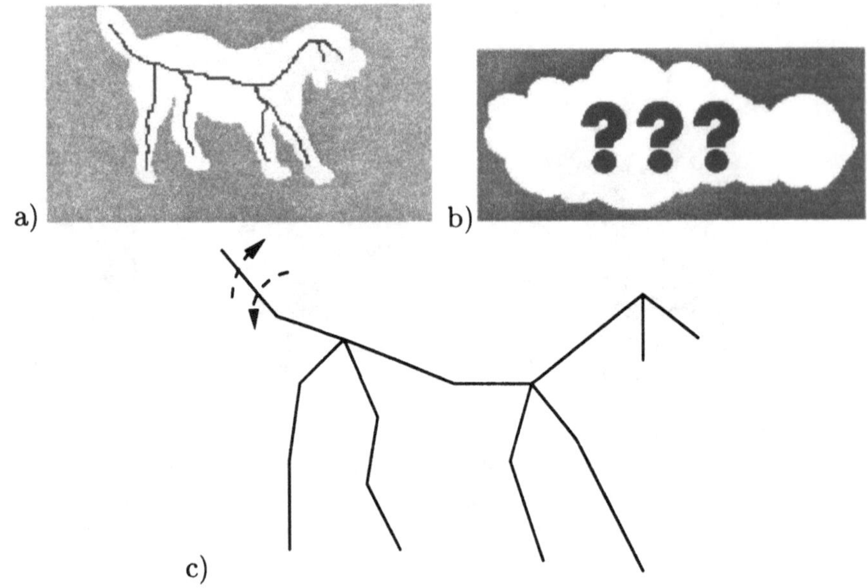

Figure 6.3. Implicit structure: a) an outline of a dog has implicit structure from its skeleton, b) an outline cloud does not follow our assumption of implicit support structure, an application of our extraction algorithm would result in noise c) a representation of a dog wagging its tail, a "happy" dog.

linked to physical properties of the objects. The extraction process of our shape query system has four components: 1) the assumption of implicit support structure, 2) the expression of these assumptions via VOS, 3) the representation of VOS with a graph, and 4) the orders enforced within the graph. The VOS can be expressed as an ordered tree, but ordered trees have inherent robustness problems that motivates our novel data structure, DAG-Ordered Trees, as a solution in Section 4..

THE ASSUMPTION OF IMPLICIT SUPPORT STRUCTURE

Through *a priori* knowledge, we can relate the object shape contour to physical properties of the object. We can use the physical properties as the basis of a shape descriptor. As shown in Figure 6.3, we can induce support structure from a given shape that acts like skeleton for the object. Our assumptions about the physical object are as follows:

1. **Support Structure Existence**
   If an object protrudes into space, we assume that a time-invariant internal support structure exists that allows this protrusion (see Figure 6.3a).

2. **Support Structure Classification**
   We assume this support structure is an invariant characteristic over the object class.

3. **Support Structure Placement**
   We assume the placement of this support structure is consistently in the middle of protrusions.

4. **Support Structure Strength**
   We assume the support structure is valued along its points and its value at a point is dependent upon the percentage of the shape that the point must support, i.e., in physical terms, the load of the object that the point must bear.

If an object follows this set of assumptions of support structure existence, placement, classification and strength, we say the object has an *implicit support structure* and this *implicit support structure* can be a representation of an object. For instance, since a skeleton does not change as the human body moves, we can use this support structure to classify unknown video objects as human. We can define a similarity measure by comparison of their support structures. To refine our similarity measure, we can place other descriptions upon our support structure, e.g., people have the same skeleton, but some can be fat and others can be thin.

Many objects do not have an implicit support structure. If an object does not have a physical skeleton (e.g., a pool of water, a blob of gelatin) or has an outline that does not give any hints about its internal structure (e.g., a house, a car), then its implicit support structure may vary over the class and may not be a good representation for the class.

For objects that follows these assumptions, the shape descriptor based upon implicit support structure allows a novel functionality: the invariance of similarity measure to bending at joints, i.e., where multiple protrusions meet. Such a shape descriptor can both recognize the person over a range of motion and become a compact representation of a person waving his hand. By annotating such a shape descriptor that is based upon this implicit support structure, we can easily describe a person walking or a dog wagging its tail.

EXPRESSION OF IMPLICIT SUPPORT STRUCTURE VIA VOS

From Chapter 3, Voronoi Order Skeleton (VOS) expresses our assumptions of implicit support structure in a mathematical form. The VOS corresponds to an object's physical skeleton, supporting our claims

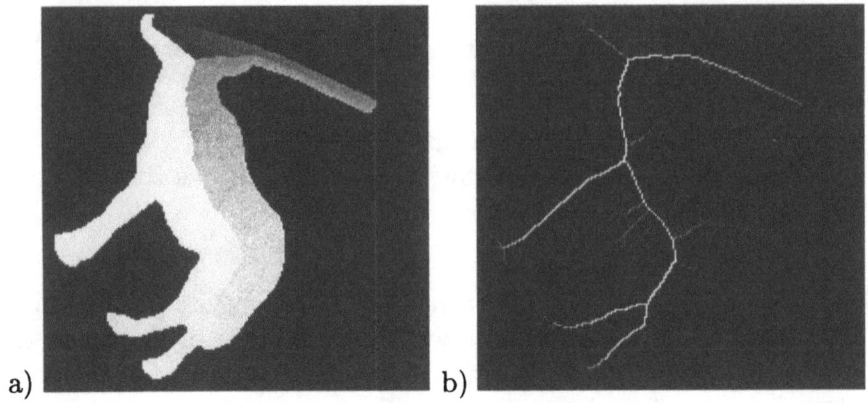

Figure 6.4. The Extraction of a Voronoi Order Skeleton (VOS): From Chapter 3, we merely calculate the Voronoi Order on the interior of the shape contour in a), and apply the Eq. 3.1 for an approximation of the VOS. Note the lightness of the image is proportional to the Voronoi Order and VOS values.

for the assumptions of support structure existence and classification. The location and values of the Voronoi Order Skeleton quantifies the concepts of support structure placement and strength, respectively.

As shown in Figure 6.4, the VOS reduces the shape to a skeleton-like structure and, for certain classes of objects, the VOS corresponds well to the projection of physical skeleton. The assumptions of the support structure existence and classification are dependent on the object class. Currently, we blindly apply VOS shape extraction to our shape contours. In the future, we would have preprocessing to recognize whether our shape descriptor is applicable to the object. It is important to note that while the physical support structure is a 3-D concept, the VOS extracts only a projection of this support structure and is also highly dependent on viewing angle of the object.

An equidistance property of the VOS formalizes our placement assumption. For the support structure placement assumption, the locations of non-zero points on the VOS are equivalent to the Medial Axis Transform (MAT). The MAT is the set of points that are equidistant from two or more points on the shape contour. Thus, the non-trivial VOS values are merely a thinned version of the original contour.

The relationship between the VOS value and the internal structure of the Voronoi Areas describes our assumption of support strength (see Section 6.). If we relate the support structure strength to the perimeter contained by the largest Voronoi Area of a point, the values of the Voronoi Order Skeleton match this support structure strength assumption.

REPRESENTING VORONOI ORDER SKELETON WITH A GRAPH

We construct graphs to represent the VOS: these graphs will be our shape descriptors and comparisons between graphs will be our similarity measure. As shown in Fig. 6.4, representation of the VOS is more compactly expressed as a composition of line segments and joints rather than as a 2-D function. We investigate the hierarchy of the VOS and transform our representation from a 2-D function to a graph.

First, we remove the majority of the VOS values by their low value. The VOS is near-zero in most areas except for selected line segments that are equidistant from two or more different points on the shape contour. Since low skeleton values correspond to low support strength, we can threshold out these values and lose only noise or unimportant features. The initial mapping of the function to the graph maintains a one-to-one correspondence between points with above-threshold values and nodes in the graph. Edges in the graph indicate that two points are adjacent and have a VOS value that is decreasing. Since the VOS values form a multiresolution tree (see Section 6.), we call this graph a *pixel-based tree* of the VOS.

To express the hierarchy as the key features of the support structure, we partition (see Section 5.3) the pixel-based tree of the VOS into composition of line segments (macro-structures) that are connected at discrete joints (breaks). We map line segments to edges and nodes to points in the skeleton where line segments meet or end. For a simple partitioning, we reduce sets of the linearly connected nodes in the pixel-based tree to an edge and form a tree. Although a tree representation describes the hierarchy of the VOS, it is sensitive to noise. To add robustness into the ordered tree we apply a complex partitioning (Section 4.) to represent the pixel-based tree.

ORDER ON THE VOS GRAPH REPRESENTATION

To apply complex partitioning, we must establish orders on this graph. Since we are representing a hierarchy on $\Re^2$ with DAGs, instead of a datastream, we require two such orders.

As shown in Figure 6.5, our ordered tree representation has two kinds of order: its depth in the tree and its order among its siblings. The position of an edge is determined by two orders: its ancestor/descendent lineage (a finger coming off an arm vs. a leg) and its siblings on the same joint (the first vs. middle finger). A tree has an order associated with its ancestor-descendant relations; for every lineage, there is a strict ordering

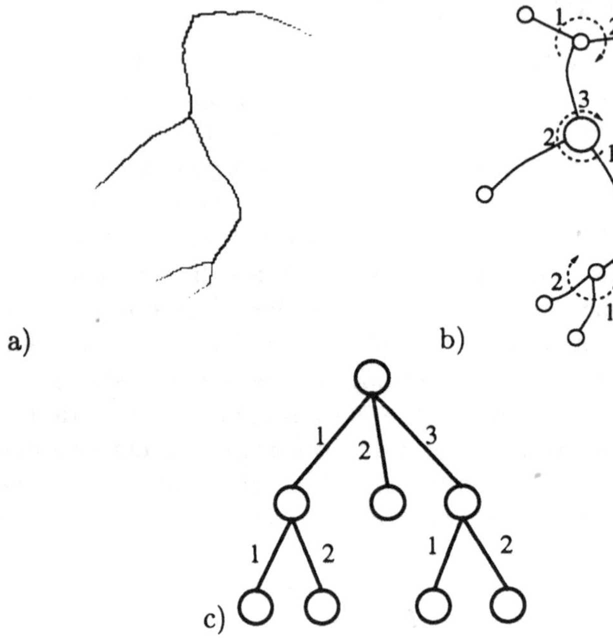

Figure 6.5. Mapping a Voronoi Order Skeleton (VOS) to a Ordered Tree: in a) we threshold the VOS values, removing noise and small shape protrusions, next, in b), we compose the image as a series of line segments joined at points, and, in c), we map the skeleton structure to ordered tree where the children are ordered by angle in the image plane.

of ancestors and descendants. An ordered tree is a tree that adds another order to the sibling relationships, i.e., between the children of the same parent. In a VOS, the sibling order and the ancestor-descendent order represent the VOS multiresolution property and the clockwise orientation of the children w.r.t. their parent, respectively. Thus, an ordered tree captures all the key features of the VOS. However, the ordered tree is sensitive to noise and partitioning errors.

4. VOS REPRESENTATION VIA DAG-ORDERED TREES (DOTS)

Although the VOS suggests an ordered tree structure, the concept of ordered trees must be generalized to DAG-Ordered Trees (DOTs, to be defined in this section) to give the necessary robustness for our shape comparison. The representation of the Voronoi Order Skeleton as an ordered tree composition of line segments may be ambiguous and/or sensitive to noise. Our solution will be based upon the DAG-based

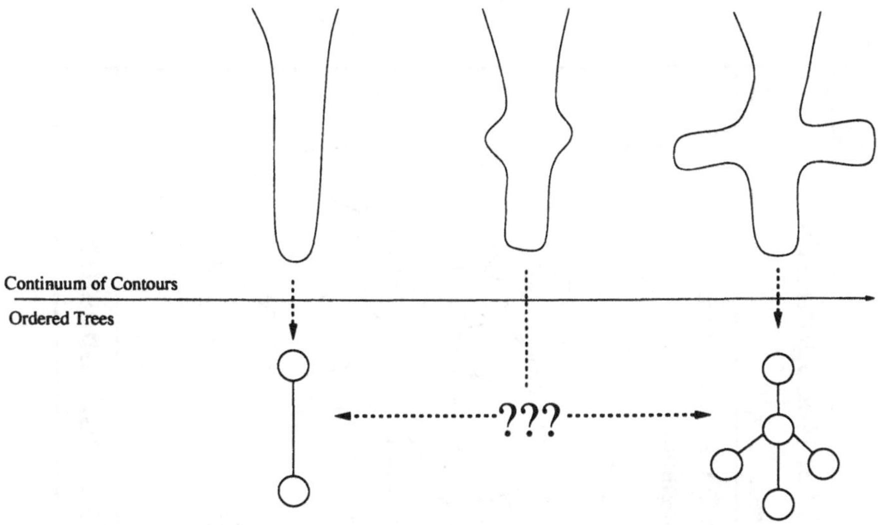

Figure 6.6. The Quantization of Hierarchy: Consider a continuum of continuous contours on the top half of the figure. From left to right, protrusions grow out of center of the line segment. At some point in the continuum, the protrusions are over the threshold, causing a large change in hierarchy of the mapped ordered tree. Although the contours are continuous, their ordered tree representations are highly sensitive to noise near the threshold. This sensitivity will be problematic in our shape comparisons.

representation described in Chapter 5 and extends the ordered tree to represent a complex partitioning of the VOS. In this section, we discuss the pitfalls of the ordered tree representation, introduce a new data structure, DAG-Ordered trees (DOTs), and motivate the DOT data structure for VOS representation. To handle a complex partitioning, we generalize the data structure of an ordered tree to a DAG-ordered tree where we order both the ancestor-descendent and sibling-sibling relationships with a DAG.

QUANTIZATION OF HIERARCHY

Whenever a continuous structure such as the VOS of a contour is mapped onto a discrete hierarchy such as a graph, *a quantization of hierarchy* affects the robustness of our ordered tree representation. Consider this simple case: let us consider a continuum of contour substructures in Fig. 6.6. At some point in the continuum, the thresholding associated with the VOS will not remove the branches, resulting in a drastic change in the ordered tree associated with the skeleton. While the contour is continuous, its corresponding ordered tree is discrete. This sensitivity within the correspondence between contours and ordered trees can cause

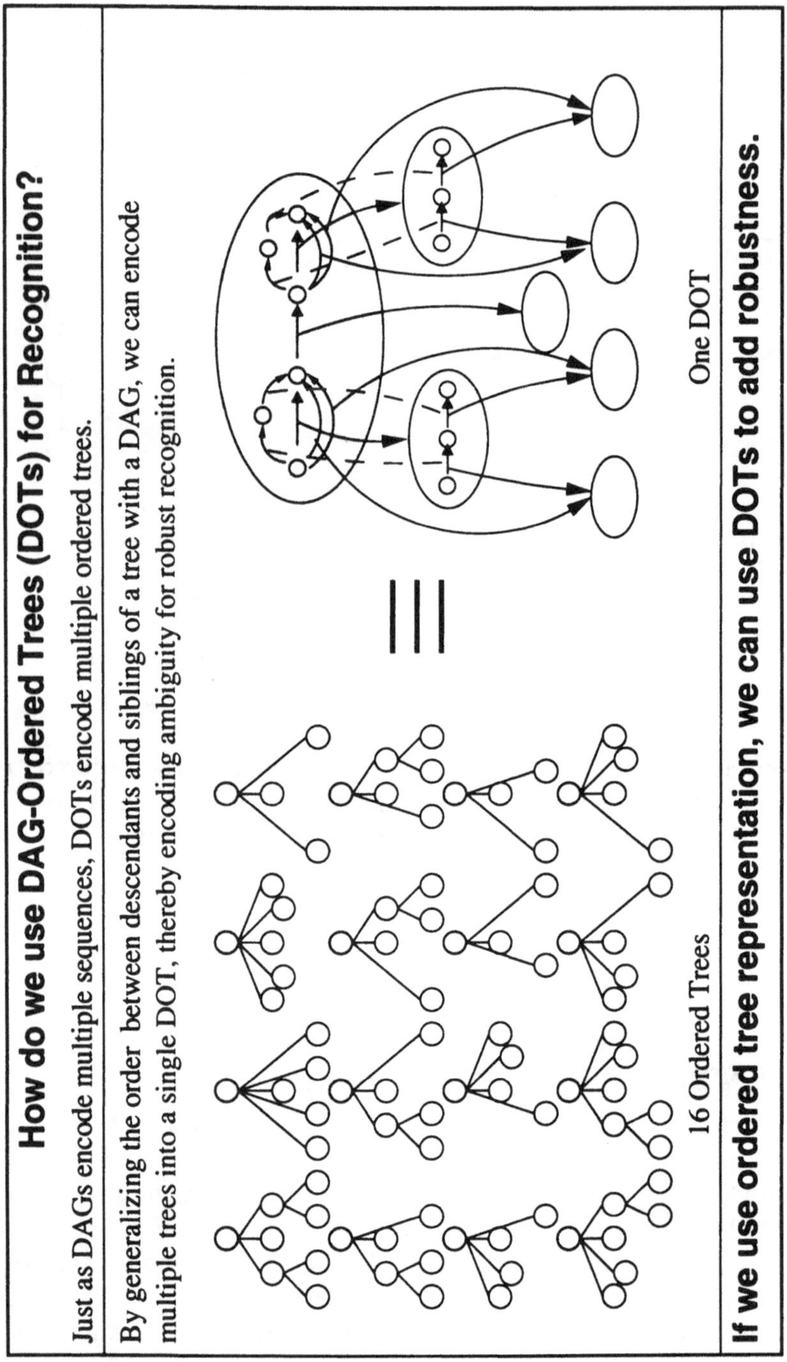
How do we use DAG-Ordered Trees (DOTs) for Recognition?
Just as DAGs encode multiple sequences, DOTs encode multiple ordered trees.
By generalizing the order between descendants and siblings of a tree with a DAG, we can encode multiple trees into a single DOT, thereby encoding ambiguity for robust recognition.
16 Ordered Trees
One DOT
If we use ordered tree representation, we can use DOTs to add robustness.

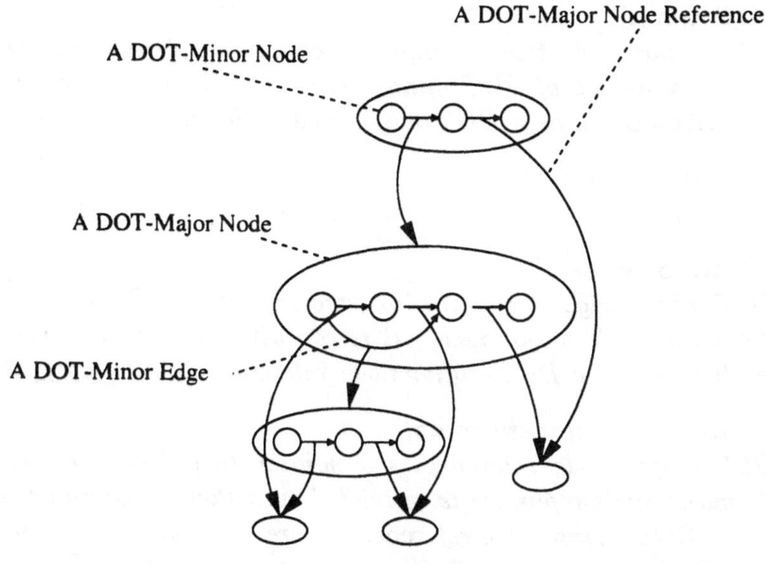

Figure 6.7. An example of a DOT with labels on terminology

many problems in matching ordered trees. When the contour is partitioned into line segments, a point of ambiguity may exist within the partitioning process at every branching point in the VOS and degrades the reliability and accuracy of the similarity measure.

We express ambiguities by merging the two ordered trees into a single data structure via complex partitioning. An ordered tree assumes one-to-one mapping of nodes and edges in the tree to line segments and joints in the VOS, respectively. However, in the presence of noise, the same contour may map to two different ordered trees. To encode these ambiguous mappings, we extend the ordered tree data structure so that both the sibling and ancestor-descendent relations are no longer lists, but are generalized to DAGs, allowing multiple ordered tree representations to better represent a continuous structure. To represent a complex partitioning of a tree, we introduce a new data structure called DAG-Ordered Trees (DOTs).

DEFINITION OF DAG-ORDERED TREES (DOTS)

DEFINITION 6.1 *(DAG-Ordered Trees (DOTs)) As shown in Figure 6.7, a DAG-Ordered tree (DOT) is a recursive directed graph structure. The component of the DOT are as follows:*

1. **the DOT-major node,**
   A DOT-major node tree is empty or contains a DAG_o (see Definition 5.1), composed of DOT-minor nodes and edges. Each edge value on the DAG_o contains a DOT-major node reference.

2. **DOT-minor node,**
   A DOT-minor node is a node contained by a DOT-major node.

3. **DOT-minor edge,**
   A DOT-minor edge is an directed edge from two DOT-minor nodes in the same DOT-major node. It is contained within a DOT-major node. It contains a DOT-major node reference and an edge value.

4. **DOT-major node reference,**
   A DOT-major node reference is a pointer to a DOT-major node. DOT-major node reference is restricted such that, if we induce graph from the DOT-major node references where 1) its nodes correspond to the DOT-major nodes and 2) its edges correspond to the DOT-major node reference, leaving the containing DOT-major node and going to the DOT-major node reference, then this graph must be a DAG.

5. **and the edge value.** *The edge values are data structures on the DOT-minor edges that represent the VOS segment and also contain a DOT-major node reference.*

Note that the ordered tree is a special case of the DOT. The DOT is an extension of the ordered tree by generalizing not only the linear sibling order to an order specified by a DAG, but also linear ancestor-descendant order to one specified by a DAG. The DOT allows the complex partitioning of the ordered tree and the efficient encoding of multiple ordered tree representations into a single data structure (see Figure 6.8). This comparison between DOTs is discussed in the Section 6.12.

5. SYSTEM DESIGN

The system implements a shape descriptor from the extraction of VOS as discussed in Chapter 3, and a similarity measure from an efficient comparison algorithm of DOT-represented support structures based on DAG work in Chapter 5. The block diagram for our shape query system is shown in Figure 6.9. In our tests, we are given a database of shapes. We extract our shape descriptors from these shapes and link their shape descriptors with their original contours. Given a query shape, we extract a shape descriptor and then calculate its similarity measures against all other shape descriptors within the database and return the top n-highest scoring shapes from the database. The next section details the two

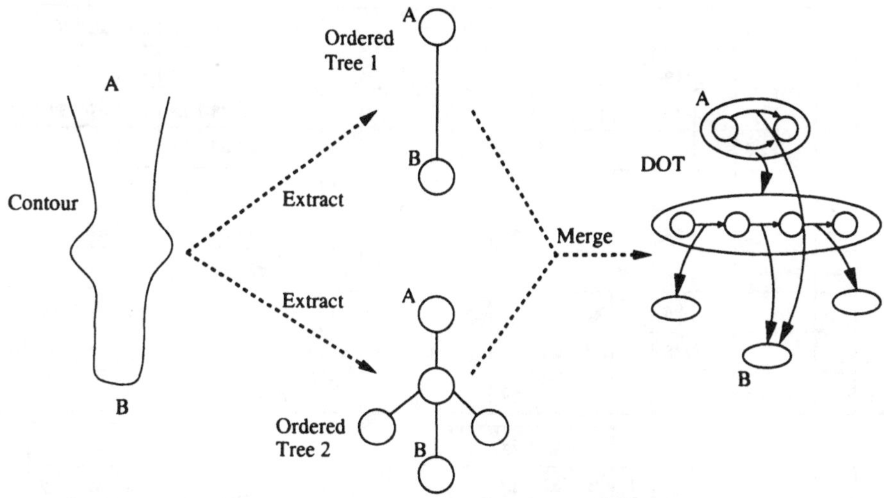

Figure 6.8. Representation of the problematic contour from Figure 6.6 can be solved by coding both ordered tree representations as one DOT and using the DOT as the shape representation.

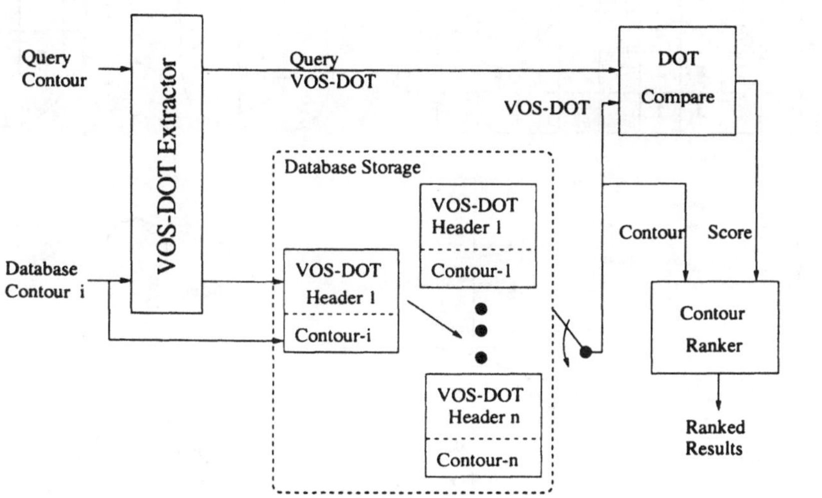

Figure 6.9. Block Diagram for Simple Shape Database

functional blocks of the system diagram: the VOS-DOT extractor that extracts our shape descriptor and the DOT-Compare that implements our similarity measure.

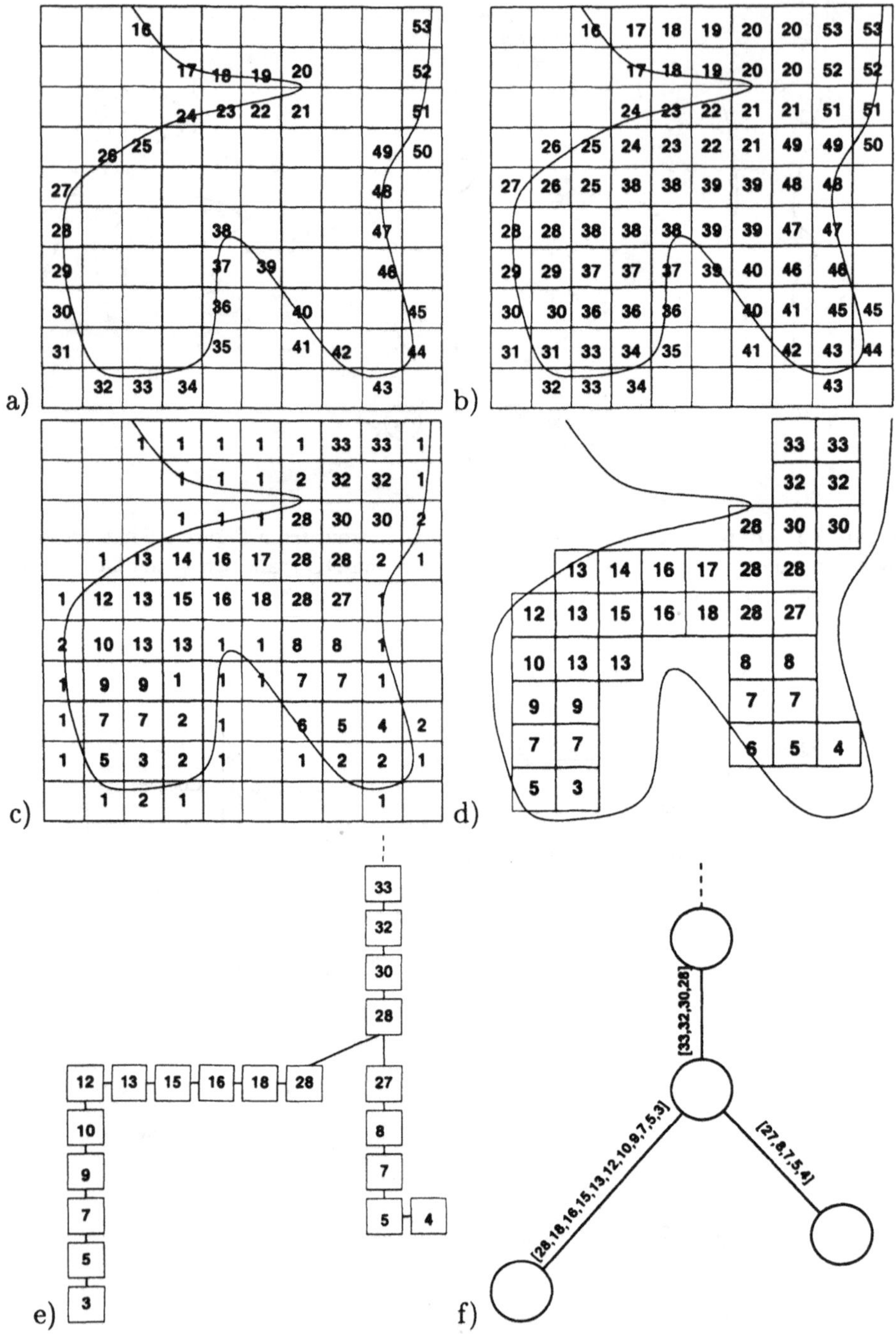

Figure 6.10. Steps in Support Structure Extraction a) Order the contour, b) Calculate the Voronoi Order, c) apply Eq. 3.2 from Chapter 3, d) threshold the pixels and map the pixels onto a graph, e) find a subgraph that is consistent with the multiresolution property, f) partition the graph into line segments.

VOS-DOT EXTRACTION

In this section, we list the steps in the extraction process that produce the shape descriptor from a given shape. The extraction process is divided into two parts: the first transforms VOS into an ordered tree and the second transforms the ordered tree into a robust shape descriptor via DOT representation.

EXTRACTION PROCESS: CONTOUR TO ORDERED TREE

The first section shows the calculation of the VOS, and uses the multiresolution property of the VOS to derive the initial ordered tree representation of the shape. After normalizing the size of the shape contour, we apply these steps, as shown in Figure 6.10:

1. **Order the contour**
   We arbitrary pick a linear order for all the points on our contour, giving an index value to every point on the contour.

2. **Calculate the Voronoi Order**
   For every point on the interior of the contour, we find the index of the closest point of the contour. The Voronoi Ordered Space can be calculated in $O(n \log(n))$, where n is the number of pixels of shape image. By mapping the grid of the image space to a graph, the Voronoi Ordered Space calculation is a modified Dijkstra's algorithm [Cormen et al., 1990] that initially places each point on the contour into the set of points of minimal distance. After the minimum distance calculations, the index of minimal distant contour point can be propagated to all pixels in $O(n)$ (where n is the number of pixels in the space).

3. **Calculate the Voronoi Order Skeleton**
   For a given point, the VOS is calculated by Eq. 3.1, i.e., maximum absolute difference of the Voronoi Order from one pixel and any of its adjacent pixels. Since the number of adjacent pixels is a constant, this step runs in $O(n)$ (where n is the number of pixels in the space).

4. **Threshold the Voronoi Order Skeleton**
   The value of the Voronoi Skeleton image is related to the support structure strength. Therefore, a simple thresholding can remove noise and small protrusions on the contour. The thresholding value of 5-10% is a good value.

5. **Find pixel-based ordered tree representation**
   After thresholding, the remaining non-zero pixels form the pixel-

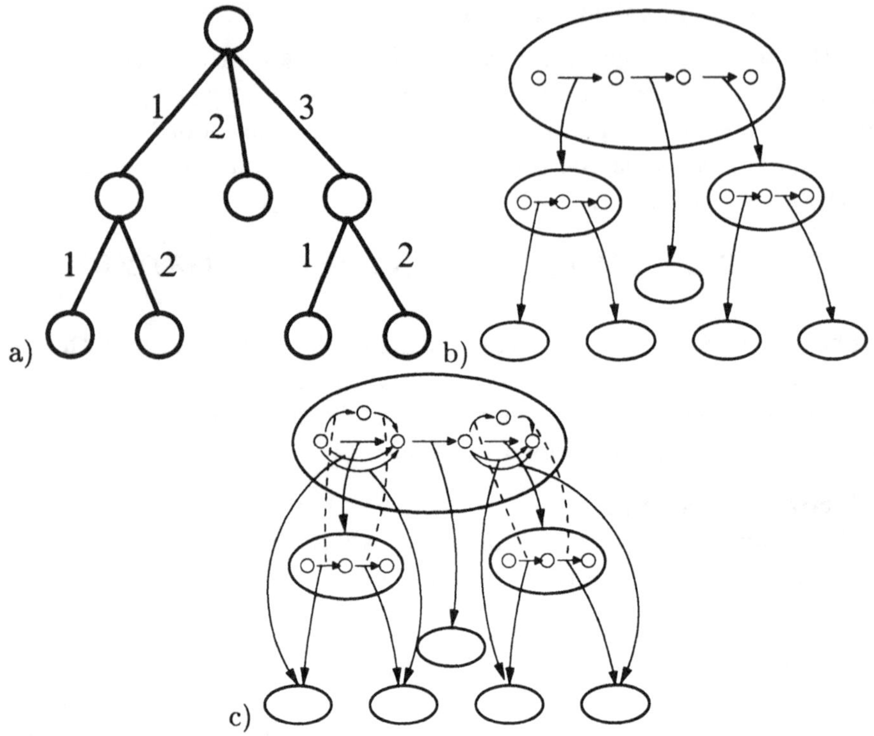

Figure 6.11. Transforming the ordered tree representation of a VOS into a robust DOT representation: a) The initial ordered tree b) Transform the ordered tree into a DAG-Ordered Tree c) Add equivalence transformations of the ordered tree structure for robustness.

based tree. To create an ordered tree, we thin this tree and extract the ordering from connectivity and VOS values. From the VOS multiresolution property, the negative gradient of values on the Voronoi Order Skeleton indicates the parent-child relationship between nodes. The order of the children of the same parent are dependent on the counterclockwise turn from the parent branch.

6. **Segment the ordered tree into line segments**
   After we derive the pixel-based VOS ordered tree, we coalesce the pixels into line segments. This operation gives us an ordered tree that will be the basis of our DOT representation.

EXTRACTION PROCESS: ORDERED TREE TO DAG-ORDERED TREE

This part of the extraction process adds the robustness into ordered tree representation through the addition of parallel structures in the

DOT. Problems in the ordered tree occur when a small perturbation in the contour may cause a large change in the hierarchy. An *extra joint* problem occurs in Fig. 6.12a where the existence of a joint is questionable. An *extra inserted branch* problem occurs when the existence of a line segment is questionable, but not its children. As shown in Figure 6.12b, if small line segment exists, then instead of having all branches hang off the same point, the branches are split in the hierarchy. Once again, a small perturbation has caused large change in the hierarchy. Although these problems may occur infrequently, the problems can occur at any node in the ordered tree. To add robustness into the ordered tree, we merely code both ordered tree representations into our DOT at every place where we recognize the possibility for extra joint and extra inserted branch problem. As shown in Figure 6.11, the steps are as follows:

7. **Transform the Ordered Tree into a DAG-Ordered Tree**
   This straightforward transformation converts each node in the tree to a DAG-Major node that contains a linear DAG_o.

8. **Add extra graph structure to DOT for robustness.**
   A robust DOT representation can be created by encoding both possibilities of each problematic case: with and without the extra inserted branch (as shown in Figure 6.12a), with and without the extra joint (as shown in Figure 6.12b). Since we are not concerned with running time, we take a particularly aggressive approach, coding each joint as a possible extra joint and each branch as a possible extra inserted branch.

9. **Extract an edge value from each line segment.**
   After our hierarchy is expressed robustly as DOT, we derive a edge values for every DOT-minor edge to represent the line segment and integrate other relevant information along with support structure into the shape descriptor.

$$e = \begin{bmatrix} length \\ curvature \\ thickness \\ VOSpercentage \\ DOTMajorRef \end{bmatrix} \tag{6.1}$$

where e is the edge value of one DOT-minor edge, and the components of e are the length of the line segment, the curvature of the line segment, the thickness of the contour at that point, the percentage of the total VOS that the line segment contains, and a DOT-major node reference, respectively.

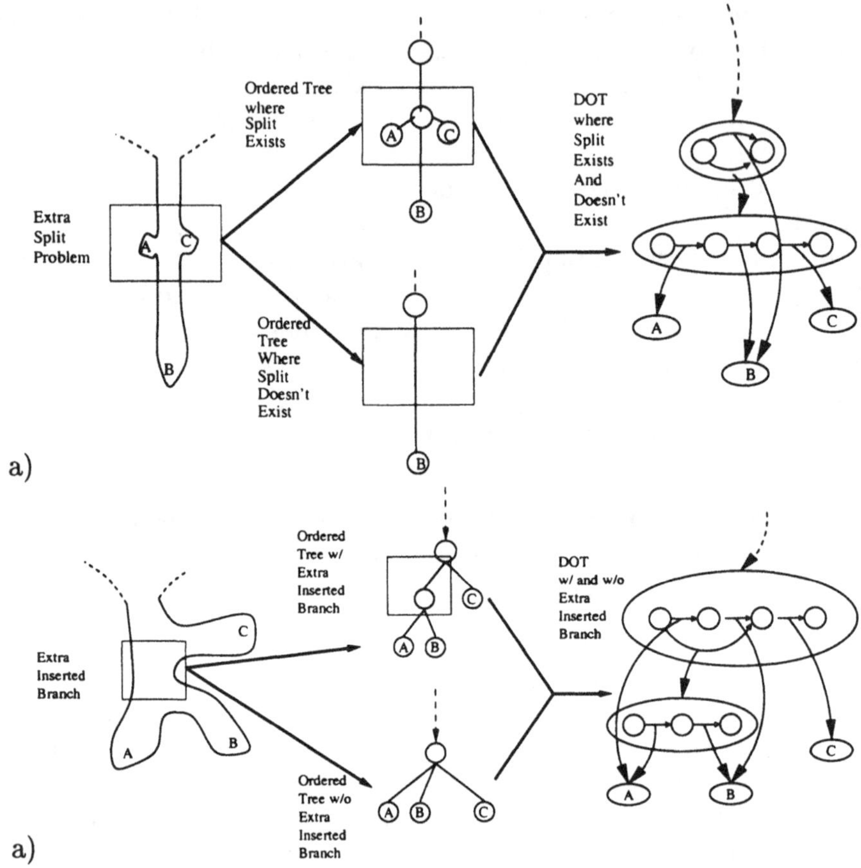

Figure 6.12. Problems of extraction and our solution in DOTs. a) the extra joint problem, the protrusions in the grayed area lead to ambiguously ordered tree extraction. Both ordered trees are coded as one DOT. b) the extra branch problem, the branch in the gray area may be inconsequential. Both ordered tree representations are coded as one DOT. Note that these problems happen at all levels of the ordered tree representation and the transformation applies to internal subgraphs as well near the leafs.

COMPARISON OF DOTS: DOT-COMPARE

To compare the DOT representations of our VOS, we use a dynamic programming algorithm called DOT-Compare, a recursive algorithm that uses the DAG-Compare algorithm of Chapter 5 in its inductive step. The DAG-Compare algorithm applies to partitioned linear datastreams and uses only the topological order among elements in the same DAG. In contrast, the DOT-Compare algorithm applies to partitioned hierarchies and, in addition to topological order of the DAG, uses the topological order in ancestor-descendent relationships to recursively compute a robust similarity measure.

By induction, we can link our DOT-Compare algorithm to the DAG-Compare operation. A path from source to sink in the DAG contained by a DAG-major node is representation of a simple partition of the multiresolution tree. The DAG-major node is compact representation of multiple simple partitions of the tree. To compare two DOTs together, we find the best matching paths between the two DAG_os of two DAG-major nodes. However, in addition to a sequence of feature vectors, every path within a DAG-major node also contains a sequence of DAG-major node references. By inductive hypothesis, we assume that all comparisons of DOTs in this sequence are solved since DOT references always point to a DOT-major node lower in the hierarchy. This operation is exactly the DAG-Compare operation as described in Chapter 5, with an MatchEdge function recursively calling the DOT-Compare algorithm.

To implement the comparison algorithm, we use the DAG-Compare algorithm with recursive calls that are short-circuited if they have previously calculated or have reached our basis (see Figure 6.13). Following a general form for DAG-Compare in Section 5., we embed a recursive DOT-Compare call in our MatchEdge operation.

$$MatchEdge_{DOT}(e_1, e_2, x) = \begin{cases} x - \text{NWeight}(e_2, \text{GetREF}(e_2)), e_1 = \lambda \\ x - \text{NWeight}(e_1, \text{GetREF}(e_1)), e_2 = \lambda \\ x - \gamma(e_1, e_2), \text{otherwise} \end{cases} \tag{6.2}$$

$$\text{where } \gamma(e_1, e_2) = \begin{pmatrix} \text{NWeight}(e_1, e_2) \cdot \\ \left(1 - \exp\left((e_1 - e_2)^T C(e_1 - e_2)\right)\right) + \\ \text{NWeight}(\text{GetREF}(e_1), \text{GetREF}(e_2)) \cdot \\ \text{DOT-Compare}(\text{GetREF}(e_1), \text{GetREF}(e_2)) \end{pmatrix},$$

where e_1 and e_2 are edge values, λ is an empty edge value, the function GetREF returns the DAG-major node reference in the edge value, $C = diag(c_{length}, c_{curvature}, c_{thickness}, c_{VOSweight}, 0)$, NWeight returns the percentage of the total VOS value that is contained by two given structures, and DOT-Compare$_{DOT}$ approximately returns the probability that the two DOTS are similar. We know that, for a given DOT-major node, all DOT-major node references that point to DOTs lower in the DOT hierarchy and we can build the comparisons up the DOT hierarchy. We must only do a quadratic number of recursive DOT-comparisons to compare the two DOTs.

The running time of the algorithm is $O(mn)$ where m is the total number of edges and nodes (both DAG-major and DAG-minor) in the

```
DOT-Compare(DOT-Major-Node A,B)
{
if (Empty(A)==0 and Empty(B)==0) // Basis #1
   return 1.0
if (Empty(A)==0 or Empty(B)==0) // Basis #2
   return 0.0
if (GetDAG(A),GetDAG(B) have been compared before {
   return (saved value)
   } else {
   return (DAG-Compare_DOT(GetDAG(A),GetDAG(B)));
   }
}
```

Figure 6.13. DOT-Compare compares two DOTs against each other. GetDAG returns the DAG_o of the given DAG-major node. There are two sets of recursion: one that is present in the code above, and the other in the function call to DAG-Compare_{DOT}, that uses a helper function in Eq. 6.2 that recursively calls DOT-Compare only on the DOTs lower in the DOT hierarchy, using the dynamic programming principle for efficient computation.

first DOT, and n is the similar value for the second DOT, since we can construct an equivalent representation for the DOT structure as a single flattened DAG. This construction is constant factor larger than the original DOT, so the running time are equivalent in the O notation.

ROTATIONAL INVARIANCE

A rotationally invariant shape descriptor can be created by coding in representations of rotated versions of the contour. To derive these rotated version, we start the topological order of the DAG_o in the DOT-major root node at every line segment that comes off of the VOS root. In this process, since the DOT structure below the root node remains the same and we add DAG-minor edges to the only top-level DOT-major node, the added complexity is proportional to the number of DOT-minor edges that we add. Thus, a comparison with rotational invariance can be computed with the complexity of $O((m + v^2)(n))$ where v is number of DOT-minor edges at the top level DOT-major node.

6. RESULTS AND ANALYSIS

Results from first small database demonstrate the functionality of our VOS-DOT similarity measure; results from the second large database show the drawbacks and trade-offs that balance our functionality. Our results confirm the system functionality and show the system is competitive with established methods of shape descriptors such as curvature space, Zernike moments and Fourier descriptors. Although the interpretation of such results is always subjective with content-based technologies, we wish to claim that our shape query systems has good enough results to validate its existence, while demonstrating its novel functionality.

DATABASE #1: SELECTIONS FROM ANIMAL DATABASE

The first test database is a small database of selected 12 animals to demonstrate functionalities and the future possibilities of VOS-DOT shape query system. This small database is biased toward our system due to the fact that most animals follow our assumptions of implicit support structure.

Our novel functionality is validated through these results upon the small database. From results in Figure 6.14, the first query (α) matches with the shape A, even though the query is rotated in 3-D space. Fortunately, in this case, the 3-D rotation does not affect the projection of physical skeleton and the similarity measure matches the two descriptors well. The second query (β) matches well with H and clearly rejects all other animals due to the six-legged structure of the beetle. For an unknown animal query (γ, an impala), our similarity measure finds the closest match to the horse shape through the removal of the horns. Finally, although the jumping horse (δ) has very different shape than (D), the skeletons remain constant and the similarity measure matches them well. From this example, the skeleton separates motion and structural information, allowing the user to annotate motion on its shape descriptor. Second and third matches can be explained through deformations of structure, intuitive (C, δ) and non-intuitive (D, α), or penalized removal of extremities (F,β). In this experiment, the situations were obviously controlled in our favor. Our next database has a more objective criterion for correctness and has a greater range of shapes to show how the VOS representation aliases and in which circumstances this shape descriptor is advantageous and disadvantageous.

DATABASE #2: MPEG-7 CORE EXPERIMENT ON SHAPE

The second large database of approximately 1400 shapes shows the effects of aliasing and an intuition about the behavior of our similarity

Shapes Database ⇒ Query ⇓	(A) 	(B) 	(C) 	(D) 	(E) 	(F) 	(G) 	(H) 
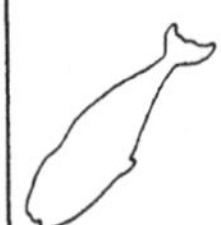 (α)	**0.77**	0.22	0.30	**0.53**	0.29	0.16	**0.31**	0.05
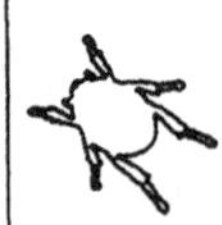 (β)	**0.28**	0.09	0.08	0.08	0.13	**0.27**	0.17	**0.47**
 (γ)	0.19	**0.30**	0.15	**0.44**	0.22	**0.24**	0.04	0.12
 (δ)	**0.42**	0.27	**0.39**	**0.53**	0.24	0.06	0.24	0.19

Figure 6.14. Ranked choices of preliminary shape query (α-δ) test on a small shape database (A-H) using VOS and DOT as representation and the DOT-Compare as similarity measure; note database and query shapes are all distinct. See Section 6.14 for analysis.

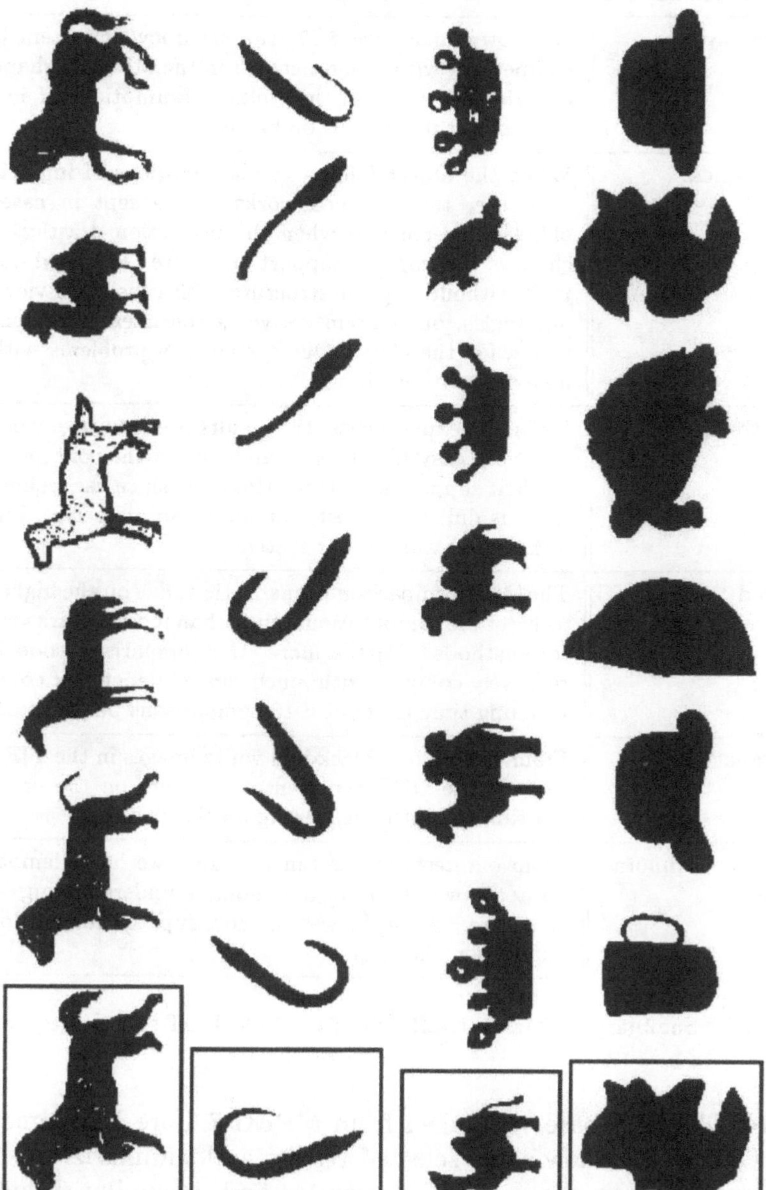

Figure 6.15. Results over a larger database; boxed shape is the query at the bottom, ranked results go from bottom to top. See Section 6.14 for analysis Top three results are bold-faced.

Quality	Comment
Accuracy	As shown in Figure 6.17, the accuracy of system is competitive with other methods in the MPEG-7 shape experiment. Objects that follow assumptions of implicit structure tend to do better.
Robustness	When the object follows our assumptions of implicit structure, the methods works well except in cases of self-occlusion and when the projection drastically changes the implicit support structure. For rigid objects without implicit structure with consistent viewing angles, our system derives a consistent representation for the class. Our system has problems with non-rigid without implicit structure.
Controllability	From our experiments, the results are intuitive, since it is very easy for users to understand the concept of implicit support structure. However, since user refinement is difficult to test, it remains an attractive, but unknown feature of our system.
Speed	The DOT comparison is quadratic and is on the higher order of the computational time than most feature vector methods. Furthermore, the comparison code is relatively complex with much control dependent code. Running time is about 6-10 comparisons per second.
Compactness	From the original black and white images in the TIFF format, the DOT representations were on the order 2-5 times smaller, depending on the shape.
Motion Annotation	From our tests of the functionality, we have demonstrated toward mimicking human understanding of structure. As such, we can now support for motion annotation shape query.

Figure 6.16. Summary of Qualitative Results from VOS-DOT shape query system.

measure. The database was taken from MPEG-7 Core Experiment on Shape Query. We show some selected results and summarize the qualitative performance of this shape descriptor and its similarity measure in Figure 6.16. However, the size of the database does not allow for example by example analysis and we compare results with the objective measure that was used in the MPEG-7 Core Experiment on Shape.

As shown in Figure 6.15, we choose a small subset of results to demonstrate the qualitative functionalities and problems of our shape query system over a large database. In the first column on the left, the query of a dog shape finds other dogs, but mostly other animals that have four legs. The second column shows the flexibility of the shape query algorithm, allowing the shape to deform while still maintaining a robust similarity measure. The third column shows the overreliance of the similarity criterion on the implicit support structure of the shape. To match the protrusions of the two-humped camel of the third column, our query system finds other shapes with just as many protrusions, rather than one-humped camels. We can see that structure takes priority over the actual shape of the animal. It is interesting to note that only two-humped camels score very well, though. The fourth column shows results of a shape that does not follow our implicit structure assumptions. As expected, the query brings rather noisy results.

For the objective measures of MPEG-7 core experiments, there were three different parts of experiment designed to test different qualities of the shape description system. Part A tested the rotation and scale invariance of the shape descriptor. Database A is comprised of 60 different images, each from a different class. To test rotational invariance, each image is copied into the database at five different rotations of 9, 36, 45, 90 and 150 degrees; the original image is then used to search the database for these 5 rotations along with the original image. A query with the original image is done and the top six choices are matched to whether they were derived from the original image. The percentage of this matching is returned. A similar test is done for scale invariance, adding scaled images of the database images at scales of 0.1, 0.2, 0.25, 0.3 and 2. The result of Test A is the average of scale and rotational invariance tests. Test B tests the classification abilities of the shape descriptor. Test B has a database of 60 different classes with 20 images per class. For every image of each class, the top 40 images are found. The final score of Test B is the average percentage of images that are from the same class that are found in the top 40. Part C tested the tolerance to video motion of the shape descriptor. Outlines from a 200 consecutive frames of video sequence of a Bream fish is mixed with 1100 outlines of other fish. The final score of Test C is the percentage of the top 200 images in the query that are of the Bream fish.

Our comparative results are given in Figure 6.17. While our results are not the best, we feel that the method is competitive with current state of the art techniques that were presented in MPEG-7 Core Experiment on Shape.

7. CONCLUSION

The MPEG-7 standard wishes to form a representational language for video objects to support the foundation of video object search engines.

Technique	Description	Test A (Size/Rot.)	Test B (Classes)	Test C (Video)
Proposal 320	Curvature Scale-Space	96	75	95
Proposal 687	Zernike Moment-based	96	70	94
Proposal 567	Wavelet-based	93	68	93
Proposal 517	Multi-layer Eigen Vector	84	71	80
Our DOT-VOS	Voronoi Order Skeletons represented by DAG-Ordered Trees	**85**	**60**	**83**

Figure 6.17. Our results from object measures in comparison to other systems from MPEG-7 Core Experiments for Shape from MPEG-7 February 1999 Meeting in Lancaster, England. See Section 6.14 for results.

Since video objects have no well-defined objective qualities, the MPEG-7 syntax supports many different types of description schemes (DS) that allow the user to search for video objects efficiently, one of which is based upon shape. This chapter introduces a shape descriptor for video objects based on assumptions of implicit support structure as the feature of a object, i.e., the skeletal structure implied by a object's image boundary. We use robust representation and comparison of these skeletal structures as a shape similarity measure via extraction by VOS and representation by DOTs. This shape descriptor is competitive with the state of the art technologies and also implements new functionalities such as invariance to bending at joints and motion annotation. We can use this shape query system to classify an unknown object's skeleton in MPEG-7 database or to search MPEG-7 database for video objects with the similar skeleton structure.

To conclude, Chapter 7 discusses the implications of extraction and representation system of the work and its future of computing with content-based information processing.

Chapter 7

THE FUTURE OF CONTENT-BASED VIDEO PROCESSING

To me, the great hope is that now these 8mm video recorders come out, just some people who wouldn't normally make movies are making movies. Suddenly, one day, some little fat girl in Ohio is going to be the next Mozart and make a beautiful film with her father's 8mm. For once, the real professionalism of movies will be destroyed and it will become an art form.
– Francis Ford Coppola

This research extends the domain of computer functionality from data to content via the technologies of content-based information processing (specifically, video processing). Currently, computers support the tasks of transporting and processing data: computing functions or doing repetitive regimented tasks. In the future, we foresee computers as experts at manipulating and processing *content*, making intelligent decisions on presenting, saving, searching or processing content. We have presented two distinct systems for video object extraction and representation in support of MPEG-4 and MPEG-7 standards, respectively. These systems are the first steps in the theoretical and algorithmic infrastructure for content-based information processing. Content-based information processing enables computers to recognize and sift content and, consequently, to maximize all resources w.r.t. content: storage, computation, and communication.

In these sections, we propose some future applications and research areas for content-based video processing.

1. UNIVERSAL MULTIMEDIA ACCESS

Content-based information processing allows computers to optimize subjective presentation of video content w.r.t. viewer resources. This application in the MPEG-7 standard is supported by Universal Multime-

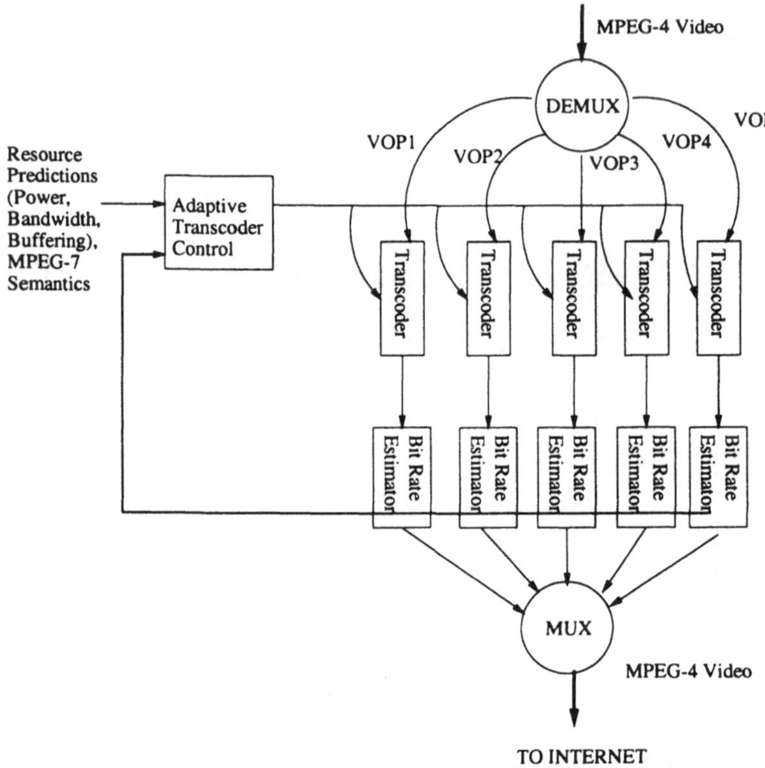

Figure 7.1. The Role of Content-Based Analysis in Universal Multimedia Access: If the encoder has access to representations of MPEG-4 sequences and knowledge of the viewer's bandwidth and hardware resources, the encoder may wish to limit the bitrates of some objects to maximize the presentation of content.

dia Access (UMA) [Mohan et al., 1998]. UMA is a part of the MPEG-7 standard that describes presentation hardware available at the viewer's end. In the spirit of MPEG-7, the syntax of UMA frames the roles of future technologies without choosing a particular implementation. UMA within the MPEG-7 syntax provides the means for the transcoder to understand the viewing resources. Given the receiver's video resources and hardware, we can recode the video stream for optimal viewing w.r.t. its content.

Given a limited resource of bandwidth or display hardware, we leverage the MPEG-4 or MPEG-7 knowledge of the video content to optimize the viewer's content (see Figure 7.1). This process of transliterating the compressed bitstream is called *transcoding.* Unlike rate control which is content-independent, transcoders recode the video to fit a given bandwidth with minimal loss of content. For instance, given a video phone

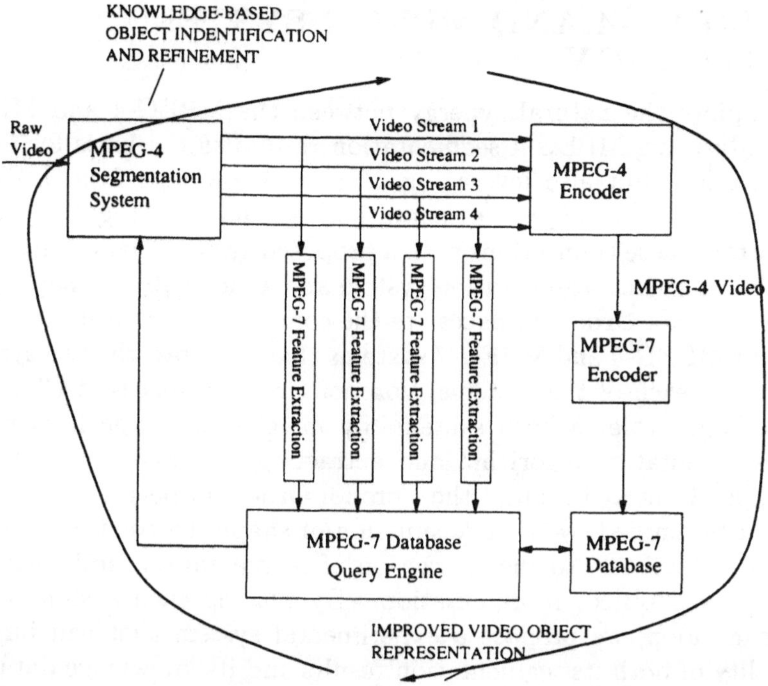

Figure 7.2. The MPEG-4 and MPEG-7 Synergy : using knowledge database, we can obtain better segmentation results. In turn, this improvement therefore can be propagated into the knowledge database.

message, the transcoder may decide not to send the MPEG-4 video object of the background of the video phone message due to the bandwidth limitations. If the bandwidth becomes further limited, the transcoder can realize that the video itself may optional and send only the audio message. If the bandwidth is even further limited or the receiver viewing hardware is a pager, the encoder may choose to send an MPEG-7 text stream. Combining the MPEG-4 object-addressable video content with the MPEG-7 content description, the transcoder has all the necessary information to optimize the presentation of content. Even when the UMA syntax is fixed in the MPEG-7 standard, there will still be challenging areas of future research: the different strageties of reducing bitrate in the compressed domain without sacrificing content and the adaptive decision process of the transcoder to optimally allocate bitrate budget w.r.t. content.

2. MPEG-4 AND MPEG-7 SYSTEM SYNERGY

We explore the natural synergy between the MPEG-4 and MPEG-7, not only using MPEG-4 segmentation results as the basis for object representation, but also leveraging *a priori* knowledge from MPEG-7 databases into video object segmentation. As shown in Fig. 7.2, we can demonstrate these connections with a proposed system that encapsulates both a video object segmentation system and a video object shape query system. Voronoi Ordered Spaces are the common theoretical thread between our MPEG-4 and MPEG-7 systems that can draw the two systems together. In each of the systems, Voronoi Ordered Spaces of Chapter 3 plays a major role in both systems by integrating shape information into a segmentation algorithm and extracting a robust representation of a shape. Centered around the Voronoi Ordered Spaces, the two systems can be unified by the transmission of shape information from the knowledge database, to the extraction of segmentation, and then back into improved object representation. By placing each system within a feedback loop, we propose a experimental system that can improve the quality of both its segmentation results and its knowledge database. Linked together in such a way, the two systems form an adaptive/neural system that begins with human guidance, but can adapt and grow on its own later on, much like the dynamic system of the human mind.

We hope to demonstrate the possibility of meta-system that learns, improving segmentation quality and, in turn, its own knowledge database. For example, we begin with an unclassified video sequence of a horse. Without any *a priori* knowledge, the segmentation system bootstraps by using low-level information such as optical flow segmentation to first determine the general blob of the horse and extracts an estimate of visual characteristics (for this system in particular, we concentrate on shape, but color, size, texture, and speed are also valid). From this first estimate, we query our video database for any archived skeletons that are consistent with the estimate. From this query, (ideally) we receive a skeleton of a horse and integrate that information to iteratively refine our segmentation result. When our segmentation result has converged, we archive a good quality segmentation in our video database and improve both segmentation and classification of future unknown sequences.

3. INTELLIGENT VIDEO TRANSPORT

In the network of the near-future (see Figure 7.3), video will be as ubiquitous as telephony, requiring Intelligent Video Transport, i.e., using content information to predict and adapt to changing network condi-

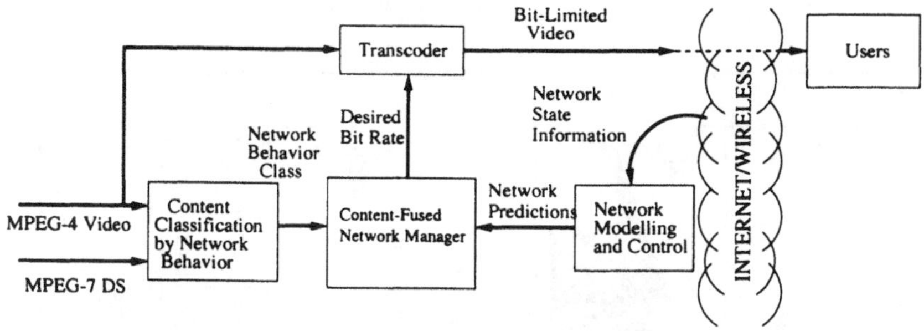

Figure 7.3. System-Level Diagram for End-to-End transmission of MPEG audio/visual data

tions[Chang and Bocheck, 1998] [Dawood and Ghanbari, 1998] [Wu and Joyce, 1999]. The major challenge of Intelligent Video Transport then is to manage large-scale wireless/mobile multimedia networks with respect to the allocation of network resources (such as buffering and bandwidth). To achieve the optimal utilization of network resources, we will research to end-to-end network transmission of MPEG video through content-based dynamic resource allocation among users and applications.

The future MPEG-7 descriptions of video content will hint at patterns of the long-term behavior of the video data. It will be vital for internet/wireless multimedia that we remap feature space of content characteristics (activity, high movement, violence, scene change, etc) onto dynamic patterns of network behavior (bit-rate, power rate, error rate, power balance). A content-based traffic model, based on relationship between video structure and its compressed representation, will be developed for the improving the prediction of the resources required by video contents. For example, face detection, sketch of trajectory/shape, indoor/outdoor detection, multiresolution model, object entry and exit may correlate well with network traffic patterns. The research will be focused on automatic classification of video sequences with respect to their network behaviors and use subjective measures of content to improve significantly video transmission quality. We will fuse MPEG-7 semantic level knowledge with network predictions and adapt to changing network conditions for best quality of service. Content behavior can be fused with current network characteristics to give more accurate forecast of network activity, and better resource allocation. Based on traffic of each content categories, the system automatically determines the optimal resource negotiation strategy for each video stream, dynamically allocating resources to improve network utilization and cope with changing network traffic.

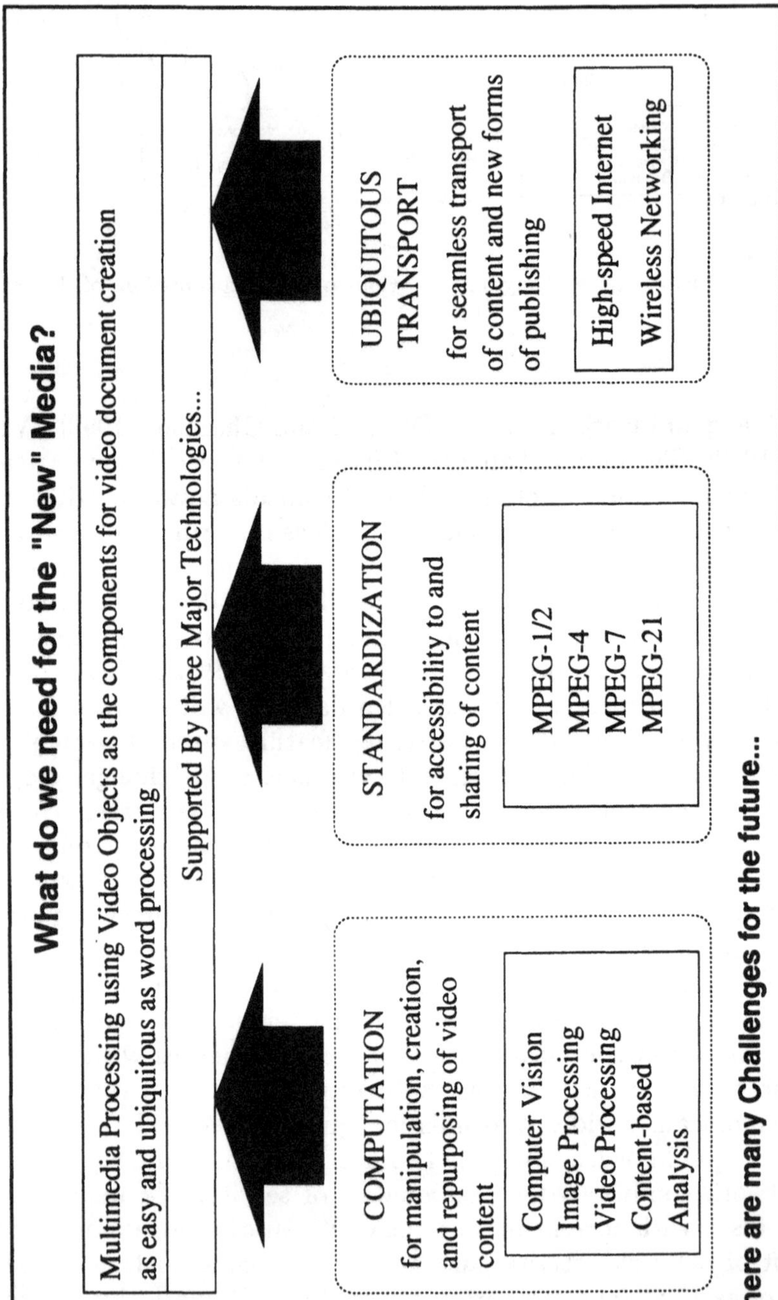
What do we need for the "New" Media?
Multimedia Processing using Video Objects as the components for video document creation as easy and ubiquitous as word processing
Supported By three Major Technologies...
COMPUTATION
for manipulation, creation, and repurposing of video content
Computer Vision
Image Processing
Video Processing
Content-based Analysis
STANDARDIZATION
for accessibility to and sharing of content
MPEG-1/2
MPEG-4
MPEG-7
MPEG-21
UBIQUITOUS TRANSPORT
for seamless transport of content and new forms of publishing
High-speed Internet
Wireless Networking
There are many Challenges for the future...

4. THE "NEW" MEDIA

To the author, the future of content-based analysis as a research field depends highly upon the existence of "new" content that is not expressible through traditional media. Video and movies are powerful presentation tools and art forms. This book advocates a movement toward video as a deeper language of communication. Let us give one example of a content-based video processing that expresses a narrative concept in a novel manner.

The movie, *The Matrix* [Wachowski and Wachowski, 1999], is a remarkable breakthrough in the video processing, because a technique of video processing actually conceptualizes a subjective reality. The notion of a moment of freezing time due to the subjective importance of an event is implemented through simultaneous video shot through a linear array of cameras on top of a synthetic background. For an event of extreme subjective important (i.e., when a bullet is shot), the notion of time also becomes subjectively important within the movie. Since simultaneous views of the same subject are taken, simultaneous frames from different cameras can be assembled into a series so that the viewing perspective moves in space, but not in time. This notion of subjective time is not represented by the subject matter of the movie, but rather by a specific technique of video processing.

In truth, we do not know where these technologies will lead or what type of video content will drive the video processing into a mainstream mode of communication. However, we do know that society and communication are intertwined: in the course of human events, people have embraced any technology that aids communication. Although we are unsure of the exact nature of the "new" media, we expect the technologies in this book and other future media-based technologies to enable new and expressive forms of communication.

References

[Adiv, 1985] Adiv, G. (1985). Determining three-dimensional motion and structure from optical flow generated by several moving objects. *IEEE Transactions on Pattern Analysis and Machine Intelligence*, PAMI-7:394–401.

[Aggarwal and Nandhakumar, 1988] Aggarwal, J. K. and Nandhakumar, N. (1988). On the computation of motion from sequences of images - a review. *Proceedings of the IEEE*, 76(8):917–35.

[Altunbasak et al., 1997] Altunbasak, Y., Oten, R., and de Figueiredo, R. (1997). Simultaneous object segmentation, multiple object tracking and alpha map generation. In *Proc of 1997 Int. Conf. on Image Proc*, pages 69–72.

[Amini et al., 1990] Amini, A., Weymouth, T., and Jain, R. (1990). Using dynamic programming for solving variational problems in vision. *IEEE Trans. on Pattern Anal. and Mach. Intelligence*, 12(9):855–867.

[Aurenhammer, 1991] Aurenhammer, F. (1991). Voronoi diagrams - a survey of a fundamental gemometric data structure. *ACM Computer Surveys*, 23:345–405.

[Bailey and Srinath, 1996] Bailey, R. and Srinath, M. (1996). Orthogonal moment moment feature for use with parametric and non-parametric classifiers. *IEEE Trans. Pattern Analysis and Machine Intelligence*, 18:389–396.

[Ballard and Brown, 1982] Ballard, D. H. and Brown, C. M. (1982). *Computer Vision*. Prentice-Hall Inc.

[Barron et al., 1994] Barron, J., Fleet, D., and Beauchemin, S. (1994). Performance of optical flow techniques. *Int'l Journal of Computer Vision*, 1(12):43–77.

[Bellegarda et al., 1994] Bellegarda, J., Nahamoo, D., Nathan, K. S., and Bellegarda, E. J. (1994). Supervised hidden markov modeling for on-line handwriting recognition. In *Proceedings of the IEEE Int. Conf. on Acoustics, Speech, and Signal Processing.*

[Bellman and Dreyfus, 1962] Bellman, R. and Dreyfus, S. (1962). *Applied Dynamic Programming.* Princeton University Press.

[Bhaskaran and Konstantinides, 1995] Bhaskaran, V. and Konstantinides, K. (1995). *Image and Video Compression and Standards: Algorithms and Architectures.* Kluwer Academic Publishers.

[Binford, 1982] Binford, T. O. (1982). Survey of model-based image analysis systems. *International Journal of Robotics, research*, 1(1):18–64.

[Blum, 1973] Blum, H. (1973). Biological shape and visual science (part i). *J. Theoretical Biology*, 38:205–287.

[Blum and Nagel, 1978] Blum, H. and Nagel, R. (1978). Shape description using weighted symmetric axis features. *Pattern Recognition*, 10:167–180.

[Bouman and Liu, 1991] Bouman, C. and Liu, B. (1991). Multiple resolution segmentation of textured images. *IEEE Transactions on Pattern Analysis and Machine Intelligence*, 13(2):99–113.

[Bouthemy and Francois, 1993] Bouthemy, P. and Francois, E. (1993). Motion segmentation and qualitative dynamic scene analysis from a image sequence. *Int. J. Computer Vision*, 10(2):157–182.

[Canny, 1986] Canny, J. (1986). A comptuational approach to edge detection. *IEEE Transactions on Pattern Analysis and Machine Intelligence*, PAMI-8(6):679–98.

[Cantoni and Carrioli, 1981] Cantoni, V. and Carrioli, L. (1981). Structural shape recognition in a multi-resolution environment. *IEEE Transactions on Pattern Analysis and Machine Intelligence*, 13(2):111–122.

[Chang and Bocheck, 1998] Chang, S.-F. and Bocheck, P. (1998). Content-based vbr traffic modeling and its applications to dynamic network resource allocation. Columbia University Technical Report 48c-98-20.

[Chen et al., 1991] Chen, L.-G., Chen, W., Jehng, Y., and Church, T. (1991). An efficient parallel motion estimation algorithm for digital image processing. *IEEE Transactions on Circuits and Systems for Video Technology*, 1(4):378–385.

[Cheng and Moura, 1999] Cheng, J.-C. and Moura, J. M. F. (1999). Capture and representation of human walking in live video sequences. *IEEE Trans. On Multimedia*, 1(2):144–56.

[Chiariglione, 1997] Chiariglione, L. (1997). Mpeg and multimedia communications. *IEEE Transactions on Circuits and Systems for Video Technology*, 7(1):5–18.

[Choi and Kim, 1996] Choi, J. G. and Kim, S. D. (1996). Multi-stage segmentation of optical flow field. *Signal Processing*, (54):109–118.

[Choi et al., 1995] Choi, J. G., Lee, S.-W., and Kim, S.-D. (1995). Segmentation and motion estimation of moving objects for object-oriented analysis-synthesis coding. In *Proc. of ICASSP95*, volume 4, pages 2431–4.

[Choi et al., 1997a] Choi, J. G., Lee, S.-W., and Kim, S.-D. (1997a). Spatio-temporal video segmentation using a joint similiarity measure. *IEEE Transactions on Circuits and Systems for Video Technology*, 7(2):279–286.

[Choi et al., 1997b] Choi, J. G., Lee, S.-W., and Kim, S.-D. (1997b). Video segmentation based on spatial and temporal information. In *Proc. of ICASSP97*, volume 4, pages 2661–2264.

[Committee, 1998] Committee, M. (1998). Mpeg-4 requirements doc., iso/iec jtc1/sc29/wg11 coding of moving pictures and associated audio mpeg98/w2194.

[Cormen et al., 1990] Cormen, T. H., Leiserson, C. E., and Rivest, R. L. (1990). *Introduction to Algorithms*. The MIT Press.

[Cunn et al., 1997] Cunn, Y. L., Bottou, L., and Bengio, Y. (1997). Reading checks with multilayer graph transformer networks. In *Proceedings of ICASSP 1997*, volume 1, page 151.

[Dawood and Ghanbari, 1998] Dawood, A. M. and Ghanbari, M. (1998). Mpeg video modelling based on scene description. In *Proceedings of Int'l Conference on Image Processing*, volume 2, pages 351–5.

[Denny, 1994] Denny, M., editor (1994). *Retina and Vitreous: Basic and Clinical Science Course Section 12.* American Academy of Ophthalmolgy.

[Duda and Hart, 1973] Duda, R. O. and Hart, P. E. (1973). *Pattern Classification and Scene Analysis.* John Wiley and Sons.

[F. and A.K, 1992] F., M. and A.K, M. (1992). A theory of multiscale, curvature-based shape representation for planar curves. *IEEE Transactions on Pattern Analysis and Machine Intelligence*, PAMI-14(8):789–805.

[Fischler and Elschlager, 1973] Fischler, M. and Elschlager, R. (1973). The representation and matching of pictorial structures. *IEEE Trans. Computers*, 1(22):67–92.

[Fu and Mui, 1980] Fu, K. S. and Mui, J. K. (1980). A survey on image segmentation. *Pattern Recognition*, 13:3–16.

[Fukushima and Imagawa, 1993] Fukushima, K. and Imagawa, T. (1993). Recognition and segmentation of connected characters with selective attention. *Neural Networks*, 6(1):33–41.

[Garcia-Salicetti et al., 1995] Garcia-Salicetti, S., Gallinari, P., Dorizzi, B., Mellouk, A., and Fanchon, D. (1995). Neural predictive approach for on-line cursive script recognition. In *Proceedings of ICASSP, IEE Int. Conft on Acoustics, Speech and Signal Processing*, volume 5, pages 3463–3466.

[Garey and Johnson, 1979] Garey, M. R. and Johnson, D. S. (1979). *Computers and Intractability: A Guide to the Theory of NP-Completeness.* W.H. Freeman and Company.

[Group, 1999] Group, R. (1999). Mpeg-7 requirements document. ISO/IEC N2727 Seoul Korea.

[Gu and Lee, 1997] Gu, C. and Lee, M.-C. (1997). Semiautomatic segmentation and tracking of semantic video objects. *IEEE Transactions on Circuits and Systems for Video Technology*, 8(5):572–84.

[Hafner et al., 1995] Hafner, J., Sawhney, H. S., and Equitz, W. (1995). Efficient color histogram indexing for quadratic form distance functions. *IEEE Transactions on Pattern Analysis and Machine Intelligence*, 17(7).

[Heitz and Bouthemy, 1993] Heitz, F. and Bouthemy, P. (1993). Multimodel estimation of discontinuous optical flow using markov random fields. *IEEE Transaction s on Pattern Analysis and Machine Intelligence*, 15(12).

[Horn and Schunck, 1981] Horn, B. and Schunck, B. (1981). Determining optical flow. *Artificial Intelligence*, 17:185–204.

[Horowitz and Pavlidis, 1976] Horowitz, S. L. and Pavlidis, T. (1976). Picture segmentation by a tree traversal algorithm. *Journal of the Association for Computing Machinery*, 23(2):368–388.

[Hotter and Thoma, 1988] Hotter, M. and Thoma, R. (1988). Image segmentation based on object oriented mapping parameter estimation. *Signal Processing*, 15(3):19–31.

[Hubel, 1988] Hubel, D. H. (1988). *Eye, Brain and Vision*. Scientific American Library.

[Hubel and Wiesel, 1979] Hubel, D. H. and Wiesel, T. N. (1979). Brain mechanisms of vision. *Scientific American*, (241):130–144.

[Irani and Peleg, 1992] Irani, M. and Peleg, B. (1992). Detecting and tracking multiple moving objects using temporal integration. In *Prof. European Conf. Computer Vision*, pages 282–287.

[Jain and Jain, 1981] Jain, J. and Jain, A. (1981). Displacement measure and its applications in interframe coding. *IEEE Transactions on Communications*, 28(12):1799–1808.

[Jeng and Tsai, 1991] Jeng, S.-C. and Tsai, W.-H. (1991). Scale and orientation-invariant generalized hough tranform – a new approach. *IEEE Transactions on Pattern Analysis and Machine Intelligence*, 24(11):1027–51.

[Kadirkamanathan and Rayner, 1993] Kadirkamanathan, M. and Rayner, P. (1993). Unified approach to on-line cursive script segmentation and feature extraction. In *Proc. of ICASSP, IEEE Intl. Conf. on Acoustics, Speech and Signal Processing*, volume 3, pages 1659–1662.

[Kass et al., 1988] Kass, M., Witkin, A., and Terzopoulos, D. (1988). Snakes:active countour models. *Int'l Journal of Computer Vision*, 1(4):321–331.

[Khotanzad and Hong, 1990] Khotanzad, A. and Hong, Y. (1990). Invariant image recognition by zernike moments. *IEEE Trans. Pattern Analysis and Machine Intelligence*, 12:489–498.

[Kung et al., 1996] Kung, S., Lin, Y.-T., and Chen, Y. K. (1996). Motion-based segmentation by principal signular vector (psv) clustering method. In *ICASSP96*.

[Kung, 1993] Kung, S. Y. (1993). *Digital Neural Networks*. Prentice-Hall.

[Lai and Vemuri, 1995] Lai, S. H. and Vemuri, B. C. (1995). Robust efficient algorithms for optical flow computation. In *International Symposium on Computer Vision*, pages 455–460.

[Lee, 1982] Lee, D. (1982). Medial axis transformation of a planar shape. *IEEE Transactions on Pattern Recognition and Machine Intelligence*, 4(4):363–9.

[Lee and Blake, 1999] Lee, S.-H. and Blake, R. (1999). Visual form created solely from temporal structure. *Science*, 284(5):1165–8.

[Legge, 1976] Legge, G. E. (1976). Sustained and transient mechianisms in human vision: Temporal and spatial properties. *Vision Research*, 18:69–81.

[Leymarie and Levine, 1993] Leymarie, F. and Levine, M. D. (1993). Tracking deformable objects in the plane using an active contour model. *IEEE Trans. on Pattern Anal. and Machine Intelligence*, 15(6):617–634.

[Lin et al., 1997] Lin, H.-C., Wang, L.-L., and Yang, S.-N. (1997). Color image retrieval based on hidden markov models. *IEEE Transactions on Image Processing*, 6(2):332–339.

[Lin and Kung, 1997a] Lin, I.-J. and Kung, S. (1997a). Coding and comparison of dags as a novel neural structure with applications to on-line handwriting recognition. *IEEE Transactions on Signal Processing*, 45(11):2701–8.

[Lin and Kung, 1997b] Lin, I.-J. and Kung, S. (1997b). A recursively structured solution for handwriting and speech recogntion. In *Proceedings of 1997 IEEE Workshop on Multimedia Signal Processing*, pages 587–592.

[Lin and Kung, 1998a] Lin, I.-J. and Kung, S. (1998a). Circular viterbi based adaptive system for automatic video object segmentation. In *IEEE Second Workshop on Multimedia Signal Processing*.

[Lin and Kung, 1998b] Lin, I.-J. and Kung, S. (1998b). A novel learning method by structural reduction of dags for on-line ocr applications.

In *Proceedings of International Conference on Acoustics, Sound and Signal Processing (ICASSP98)*.

[Lin and Kung, 1999] Lin, I.-J. and Kung, S. (1999). Automatic video object segmentation via voronoi ordering and surface optimization. In *1999 IEEE Third Workshop on Multimedia Signal Processing*.

[Lin et al., 1998] Lin, I.-J., Vetro, A., Sun, H., and Kung, S.-Y. (1998). Circular viterbi: Boundary detection with dynamic programming, mpeg98/doc. 3659.

[Lin et al., 1996] Lin, Y.-T., Chen, Y.-K., and Kung, S. (1996). Object-based scene segmentation combining motion and image cues. In *ICIP*.

[M. et al., 1995] M., G., Leroux, M., and Bertille, J.-M. (1995). Strategies for cursive script recognition using hidden markov models. *Machine Vision and Applications*, 8(4):197–205.

[Macleod and Rosenfeld, 1974] Macleod, I. D. G. and Rosenfeld, A. (1974). The visilibility of gratings: Spatial frequency channels or bar-detecting units. *Vision Research*, 14:909–915.

[Malladi et al., 1995] Malladi, R., Sethian, J., and Vemuri, B. (1995). Shape modelling with front propagation: A level set approach. *IEEE Transactions on Pattern Analysis and Machine Intelligence*, 2(17):158–175.

[Man and Poon, 1993] Man, G. M. and Poon, J. C. (1993). Cursive script segmentation and recognition by dynamic programming. In *Proc. of SPIE, The Intl. Soc. for Optical Engineering*, volume 1906, pages 184–194.

[Manke et al., 1994] Manke, S., Finke, M., and Waibel, A. (1994). Combining bitmaps with dynamic writing information for on-line handwriting recognition. In *Proc. of the 12th IAPR Intl. Conf. on Pattern Recognition*, volume 2, pages 596–598.

[Marr, 1988] Marr, D. (1988). *Vision*. W.H. Freeman and Company.

[Mayya and Rajan, 1995] Mayya, N. and Rajan, V. (1995). An efficient shape representation scheme using voronoi skeleton. *Pattern Recognition Letters*, 16:147–160.

[Meier and Ngan, 1998] Meier, T. and Ngan, K. N. (1998). Automatic segmentation of moving objects for video object plan generation. *IEEE Transactions on Circuits and Systems for Video Technology*, 8(5):525–538.

[Menon et al., 1996] Menon, R. P., Acharya, R., and Zhang, A. (1996). Content based image query from image databases using spatio-temporal transform and fractal analysis methods. In *Proceedings of SPIE*, volume 3, pages 863–866.

[Micheli, 1994] Micheli, G. D. (1994). *Synthesis and Optimization of Digital Circuits.* McGraw-Hill Inc.

[Mohan et al., 1998] Mohan, R., Smith, J., and Li, C.-S. (1998). Multimedia content customization for universal access. In *Proceedings of SPIE - The International Society for Optical Engineering*, volume 3527, pages 410–18.

[Mohri, 1997] Mohri, M. (1997). Finite-state transducers in language and speech processing. *Computational Linguistics*, 23.

[Nagel, 1987] Nagel, H.-H. (1987). On the estimation of optical flow: Relations between different approaches and some new results. *Artificial Intelligence*, (33):299–324.

[Naseri and Stuller, 1996] Naseri, H. and Stuller, J. A. (1996). Segmentation and motion estimation. In *Proc. of ICASSP96*, pages 1906–9.

[Neri et al., 1998] Neri, A., Colonnese, S., Russo, G., and Talone, P. (1998). Automatic moving object and background separation. *Signal Processing*, 2(66):219–232.

[Nishida, 1995] Nishida, H. (1995). Model-based shape matching with structural feature grouping. *IEEE Transactions on Pattern Analysis and Machine Intelligence*, 17(3).

[Ogniewicz and Kubler, 1995] Ogniewicz, R. L. and Kubler, O. (1995). Hierarchical voronoi skeletons. *Pattern Recognition*, 28(3):343–359.

[Ong and Spann, 1996] Ong, E. and Spann, M. (1996). Robust multiresolution computation of optical flow. In *Proceedings of ICASSP96*, volume 4, pages 1938–41.

[Pappas, 1992] Pappas, T. N. (1992). An adaptive clustering algorithm for image segmentation. *IEEE Transactions on Signal Processing*, 40(4):901–14.

[Parizeau and Plamondon, 1995] Parizeau, M. and Plamondon, R. (1995). A fuzzy-syntactic approach to allograph modelling for cursive script recognition. *IEEE Transactions on Pattern Analysis and Machine Intelligence*, 17(7).

[Prokop and Reeves, 1992] Prokop, R. and Reeves, A. (1992). A survey of moment-based technoiques for unoccluded object representation and recognition. *Graphical Models and Image Processing*, 54(5):438–460.

[Rabiner, 1989] Rabiner, L. R. (1989). A tutorial on hidden markov models and selected applications in speech recognition. *Proceedings of the IEEE*, 77(2):257–86.

[Ramadge and Wonham, 1989] Ramadge, P. J. and Wonham, W. M. (1989). Supervisory control of a class of discrete event processes. *SIAM Journal of Control and Optimization*, 25(1):206–30.

[Remias et al., 1996] Remias, E., Sheikholeslami, G., and Zhang, A. (1996). Block oriented image decomposition and retrieval in image database systems. In *Proceedings of 1996 Int. Workshop on Multimedia Database Management Systems*, pages 85–92.

[Rosenfeld, 1986] Rosenfeld, A. (1986). Axial representations of shape. *Computer Vision, Graphics, and Image Processing*, (33):156–173.

[Samal and Edwards, 1997] Samal, A. and Edwards, J. (1997). Generalized hough transform for natural shapes. *Pattern Recognition Letters*, 18:473–480.

[Schenkel et al., 1994] Schenkel, M., Guyon, I., and Henderson, D. (1994). On-line cursive script recognition using time delay neural networks and hidden markov models. In *Proc. of ICASSP, IEE Int. Conf. on Acoustics, Speech and Signal Processing*, volume 2, pages 637–640.

[Schomaker, 1993] Schomaker, L. (1993). Using stroke-or character-based self-organizing maps in the recognition of on-line, connected cursive script. *Pattern Recognition*, 26(3):443–450.

[Shapiro, 1980] Shapiro, L. G. (1980). A structural model of shape. *IEEE Transactions on Pattern Analysis and Machine Intelligence*, PAMI-2(2):111–126.

[Sikora, 1997] Sikora, T. (1997). The mpeg-4 video standard verification model. *IEEE Trans. Circuits Syst. Video Technology*, 7:19–31.

[Singer and Tishby, 1994] Singer, Y. and Tishby, N. (1994). Dynamical encoding of cursive handwriting. *Biological Cybernetics*, 71(3):227–237.

[Smith and Chang, 1997] Smith, J. R. and Chang, S.-F. (1997). Enhancing image search engines in visual information environments. In *Proceedings of IEEE First Workshop on Multimedia Signal Processing*, pages 545–552.

[The and Chin, 1988] The, C.-H. and Chin, R. T. (1988). On image analysis by the methods of moments. *IEEE Transactions on Pattern Analysis and Machine Intelligence*, 10(4).

[Til Aach, 1993] Til Aach, Andre Kaup, R. M. (1993). Statistical model-based change detection in moving video. *Signal Processing*, 31(2):165–180.

[Tsitsiklis, 1989] Tsitsiklis, J. N. (1989). On the control of discrete-event dynamical systems. *Math. Control Signals Systems*, (2):95–107.

[W. et al., 1995] W., C., Lee, S.-W., and Kim, J. (1995). Modeling and recognition of cursive words with hidden markov models. *Pattern Recognition*, 28(12):1941–1953.

[Wachowski and Wachowski, 1999] Wachowski, A. and Wachowski, L. (1999). The matrix. Produced by Joel Silver.

[Wallace, 1991] Wallace, G. K. (1991). The jpeg still picture comopression standard. *Communications of the ACM*, 34:30–44.

[Wang, 1995] Wang, D. (1995). Unsupervised video segmentation based on watersheds and temporal tracking. *IEEE Transactions on Circuits and Systems for Video Technology*, 8(5).

[Wang and Adelson, 1994] Wang, J. Y. and Adelson, E. H. (1994). Representing moving images with layers. *IEEE Transactions on Image Processing*, 3(5):625–638.

[Wang et al., 1995] Wang, M.-J. J., Wu, W.-Y., Huang, L.-K., and Wang, D.-M. (1995). Corner detection using bending value. *Pattern Recognition Letters*, (16):575–83.

[Wilson and Bergen, 1979] Wilson, H. R. and Bergen, J. R. (1979). A four mechianism model for threshhold spatial vision. *Vision Research*, 19:19–32.

[Wu and Joyce, 1999] Wu, M. and Joyce, R. (1999). Ele571 project:.

[Yeo and Liu, 1995] Yeo, B. and Liu, B. (1995). Rapid scene analysis on compressed videos. *IEEE Transactions on Circuits and Systems for Video Technology*, 5(6):533–544.

[Yu and Wolf, 1997] Yu, H.-H. and Wolf, W. (1997). A visual search system for video and image databases. In *Proceedings of 1997 Int. Conf. on Multimedia Computing and Systems*, pages 517–24.

[Zucker, 1976] Zucker, S. W. (1976). Survey: Region growing: Childhood and adolesence. *Computer Graphics and Image Processing*, (5):382–399.

Index

3-D DCT, 16
Adaptive signal processing, 10
Affine motion, 42
Aliasing
  of shape descriptor, 149
Ambiguities, 139
Ambiguity, 114
Ancestor-descendant relations, 135
Ancestor-descendant relationship, 62
Aperture problem, 35
Applicability of a model, 43
Area thresholding, 32
B frames, 16
Background extraction, 71
Block Size, 16
Block matching, 38
  problems with, 38
Boundary motion coherent, 78
Boundary-based representation, 31
Breaks, 112
CIF image, 25
Calculus of variations, 39
Canny Edge Filter, 89
Cellular specializations, 26
Chain rule, 34
Change Detection, 68
Change detection, 36, 78
Choi's work, 70
Circularly ordered space, 87
Classification technologies, 43
Clique finding, 31
Clustering, 31
Color, 70, 76
Complex partitioning, 137, 140
Computer vision, 11
Cones, 28
Connectivity, 32
Containment relation, 62
Containment, 80
Content percentage, 101
Contour optimization, 86
Coordinate system
  Cartesian, 50
  an alternative, 53
Coordinates
  polar, 89
Correspondence problem, 35
Coverage percentage, 101
Cursive handwriting recognition, 107
Curvature space, 149
Curvature-scale space, 128
Curve representation techniques, 107
Cyclic dependence, 117
DAG, 116
DAG-Coding, 21
DAG-Compare, 21, 120, 122
DAG-Ordered Trees, 22, 132
DAG-Ordered trees (DOTs), 137
DAG-ordered tree (DOT), 20
DOT-Compare, 146
Datastreams, 111
Decomposition via partitioning, 111
Depth First Search, 31
Dijkstra's algorithm, 143
Dijkstra's shortest path algorithm, 51
Discrete Cosine Transform, 15
Divide and conquer, 113, 122
Duality in problem statements, 29
Dynamic programming, 43, 74, 86
Edge detection, 31–32, 76
Edge following algorithms, 33
Edge values, 116
Eigenvector, 84
End-to-end network transmission, 159
Energy function taxonomy, 74
Equidistance property, 134
Euclidean geometry, 47
Excitory region, 29
Expectation and Maximization (EM), 94
Expectation-Maximization, 20

External energies, 75
Extra inserted branch, 145
Extra joint, 145
False positives, 103
Finite State Transducers (FSTs), 109
Fitness metric, 75
Fourier descriptors, 149
Gauss-Seidel, 40
Generalized Hough Transform, 128
Graceful degradation, 26
Gradient method, 86
Graph coloring, 31
Grassfire algorithm, 47
Hidden Markov Model, 11
Hidden Markov Models (HMM), 107
Hierarchy, 135
High ordered statistics (HOS), 37
Homogeneous regions, 37
Horn and Schunck, 39, 77
I frames, 16
Image query problem, 127
Image segmentation, 7
Image understanding, 30
Impala, 149
Implicit structure, 64
Implicit support structure, 132–133
Independent set, 31
Inductive hypothesis, 147
Inhibitory region, 29
Intelligent Video Transport, 158
Inter-object relationships, 73
Intermediate representations, 2
Internal energies, 75
Iterative Viterbi, 20–21, 90
JPEG, 15
Kalman filter, 42
Kalman filters, 79
Least Square, 38
Level sets, 50
Lighting conditions, 35
Linear orders, 109
Locality, 80
Longest Common Subsequence, 118
MPEG committee, 12
MPEG standards, 13
MPEG-1/2, 13, 15
MPEG-4 Problem Simplifications, 70
MPEG-4 test sequences, 67
MPEG-4, 13, 16, 67, 105
MPEG-7 Core Experiment on Shape Query, 152
MPEG-7, 13, 18, 127–128
Macro-structure, 111
MatchEdge function, 119
MatchEdge, 120
Media-based goal, 11
Medial Axis Transform (MAT), 47, 134
Mesh-based analysis, 68
Metaphor of the cave, 25
Minimum cut, 31
Model fitting, 33
Monotonicity
  of MatchEdge, 120
Motion Annotation, 131
Motion clustering, 77
Motion discontinuities, 78
Motion estimation, 9
Motion paths, 34, 84
Motion smoothness, 34
Motion
  as object discriminant, 33
Multiresolution representations, 22
Multiresolution tree, 58, 135
Neighborhood for block matching, 38
Neural networks, 10
Newton-Raphson, 86
Non-verbal query, 20
Object detection, 68
Object tracking, 67
Object-addressable content, 67
Object-addressable, 17, 67
Object-oriented decomposition, 4
Objective segmentation quality measure, 96
Occlusion
  problems with, 101
  robustness to, 98
Ogniewicz and Kubler, 51, 58
Optical flow, 34, 38
  counterexamples, 34
  multiresolution, 77
  problems with, 40
Optimal substructure, 44
Order, 109
Ordered mapping, 109
Ordered tree, 140
Orthonormal projections, 50
Overlapping subproblems, 44
P frames, 16
Parametric models, 41
Partitioning, 111, 113
  complex, 113
  imperfect, 114
  simple, 113
PathScore, 118, 120
Peak/saddle regions, 84
Physical objects
  definition of, 4
Physical space
  definition of, 4
Physical world
  definition of, 4
Pixel-based tree, 135
Principal Component Analysis, 83
Psychovisual experiments, 26

Pyramidal image representation, 33
Quality vector, 101
Quantization of hierarchy, 137
Query functionality, 43
Rate control, 16
Recognition technologies, 107
Recursive system architecture, 122
Reduction of dimensionality, 74
Region growing algorithm, 33
Region-based representation, 31
Retina, 26
Rods, 28
Rotational invariance, 148
Running time, 96, 147
Schwartz Inequality, 52
Shape descriptor, 130, 140–141
Shape descriptors, 135
Shape query problem description, 128
Sibling relationships, 136
Similarity measure, 130, 135, 141
Similarity measures, 140
Smoothness, 38
Snake algorithms, 33
Snakes, 68
Spatio-temporal correlation, 33
Spurs, 114
Steepest Descent, 86
Subjective presentation, 155
Sum of Squared Difference, 38
Surface optimization, 20, 73
Surface-based description, 73
Synergy between the MPEG-4 and MPEG-7, 158
Taylor's series expansion, 34
Texture analysis, 70
Texture, 76
The Matrix, 161
Thresholding value, 143
Thresholding, 135, 137
Time-dependent neural nets (TDNN), 107
Topological order, 111, 146
Transcoding, 156
Trichromaticity, 28
Universal Multimedia Access (UMA), 155
User feedback, 130
VOS-DOT extractor, 141
Video Object Query by Shape, 19
Video Object Segmentation, 19
Video object plane (VOP)
  definition of, 6
Video object segmentation, 17, 19, 70
Video object
  definition of, 6
Video objects, 127
Video sequence
  definition of, 4
Video space
  definition of, 4
Vision-based goal, 11
Visual ambiguities, 35
Visual processing, 2
Viterbi Rings, 76, 79, 87
Viterbi Trellis
  Warped, 91
Volume-based approach, 73
Voronoi Areas
  definition of, 58
  relations between, 59
Voronoi Cells, 47
Voronoi Center
  definition of, 52
Voronoi Distance, 50
Voronoi Order Consistent, 63
Voronoi Order Skeleton (VOS), 55
  approximation of, 55
Voronoi Order Skeletons (VOS), 128
Voronoi Order consistency, 80
Voronoi Order, 50
Voronoi Ordered Space, 50
  applications of, 63
Voronoi Ordered Spaces, 158
Voronoi Pivot
  definition of, 59
Voronoi Pivots
  tree structure of, 61
Voronoi Projection
  definition of, 52
Voronoi Skeletons, 127
Voronoi-containment relation, 62
Warped version of 2-D image space, 54
Watershed analysis, 68
Wireless/mobile multimedia networks, 159
Zernike moments, 149
Zero crossings, 32

SPRINGER NATURE

GPSR Compliance

The European Union's (EU) General Product Safety Regulation (GPSR) is a set of rules that requires consumer products to be safe and our obligations to ensure this.

If you have any concerns about our products, you can contact us on ProductSafety@springernature.com

In case Publisher is established outside the EU, the EU authorized representative is:

Springer Nature Customer Service Center GmbH
Europaplatz 3
69115 Heidelberg, Germany

FSC
www.fsc.org
MIX
Papier | Fördert gute Waldnutzung
FSC® C083411

Kolibri 360 GmbH
Ferdinand-Jühlke-Straße 7
99095 Erfurt, Deutschland
produktsicherheit@kolibri360.de